Pro-activity

Save costs & Optimize profit

by

Paul Potargent

authorHOUSE®

AuthorHouse™ UK Ltd.
500 Avebury Boulevard
Central Milton Keynes, MK9 2BE
www.authorhouse.co.uk
Phone: 08001974150

First published by AuthorHouse 1/14/2010

ISBN: 978-1-4490-2079-8 (sc)

This book is printed on acid-free paper.

In an increasingly competitive world, we believe
it's quality of thinking that gives you the edge
– an idea that opens doors, a methodology that
prevents problems or an insight that makes sense
of it all.

The more you know earlier,
the smarter and faster you can go.

CIP
Continuous Improvement Programs
solve
problems.

The Pro-active Programs
prevent
problems.

I hope I will be able to

- open your eyes -

- let you use the processes -

- let you interpret your work -

- let you create and add value -

- let you work pro-active -

- stimulate your creativity -

for customer's needs.

Index

You make the difference for your customer.

Your customer's expectation is your food.

Don' eat healthy

&

die.

Pro-activity

In

Business Processes

Product Creation

Manufacturing

Product Portfolio

Idea Creation and Innovation

Value Creation

Save Costs & Maximize Profit

Pro-activity 1: The profit reality
Earn money & Achieve Customer Satisfaction

You will earn money by being pro-active.

On the contrary to what others ever wrote in their books and the disappointment you got when you didn't earn money, in this case it is really true. You will earn money by doing things in a pro-active way because you don't need to repair the stupidities that you created. Pro-active means that you are well prepared, you did really your very best to get all details about the project right and you gave it not your best shot but you gave it

"YOUR UTMOST BEST SHOT"

Even if someone would challenge you that he can do your project better you are sure that you took the best possible approach with all the people you have. The challenger was not there when you made your decisions and as such the statement that he could do a better job is irrelevant. The pro-active methodology is one to avoid most mistakes that we make, that companies make, that women make, that politicians make and that everybody, except men, makes.

By avoiding the mistakes, stupidities, blunders and wrong resource assignment we create the benefit and we get the real profit during and after the process execution by
highest quality
&
assured Time To Market.

A critical reader will notice that I didn't write "that men make". Any explanation why I didn't write "that men make"? It is very easy. "Men never make mistakes". I suggest that women read further anyway because even if you would think that this is a discriminative statement about "women", it is really the opposite.

Men don't make mistakes.

Most of the managers are "men" and most of the mistakes are also made by men but on the contrary of what we should presume as normal, in most companies the managers that solve problems are the best rewarded and have career success because they effectively do something visible. They solve problems. That's true. But perhaps they created the problems too. My opinion is that the opposite should be rewarded. The managers that take care about their business in a way that there are no issues should be rewarded. The contrary is so much true.

It can also be that you apply the Dilbert principle, that you make others pay for your mistakes or that you make sure that you are far away from the events. If you don't make mistakes in a degree that you have to use the word "fire fighting" sometimes than you have to be working already in a way that has to be somehow related to pro-activity. You are already using an adaptive method to achieve your goal. Most companies have a solid process layout and products are created according to these processes, see ISO9001. However what should be done if a crucial parameter changes in the equation? Should we just run against the wall? Re-active processes will run into trouble because of the fact that Continuous Improvement Processes (CIPs) can only improve when problems occurred already.

The pro-active approach is set up to act before things go wrong.

Perhaps you have or perhaps you don't have but now you should make sure that you have the customer on top of your priorities. Did you meet the customer's expectations or did you meet your own expectations? With our pro-active approach we try to meet the customer's expectations first.

This pro-active book is created to meet the customer's expectations.

The logical consequence will be that we meet our goals too. For pro-activity I focus on avoiding mistakes and using competence at the highest level and it will have an effect on "wastes" too.

A short explanation about wastes: The 3 known wastes from Kaizen are muda (waste, a Japanese ugly word for doing something but not adding value), muri (overburdening system, pushing more through a pipe than it can handle) and mura (unevenness of flow, assure a reasonable steady pull). In these cases we don't speak about mistakes anymore but about optimization of speed, costs and price also called optimized tuning or lean sigma. Although many potential costs are calculated in the next sheets, the content of the book is not about speed, price and wastes. These numbers are only to get an idea about the potential savings when we do things right.

You will earn money.
For our cost savings estimations we use the following givens.

1.000.000 units Sales
Unit price 10$
Margin 40%

Total potential "costs – profit"

In Business	In T&D / R&D	In Factory
2.700.000$	1.181.000$	1.151.000$

Note: don't discuss about the 40% margin. I know that some companies only take 3% margin. The product turnaround time makes that your year margin is still acceptable(12/product turnaround time x 3%), I hope for you. So, pay attention in this case that you have margin on year base, ebit.

Costs-profit originated by Business

Business: Why use pro-active DFSS?

Facts:	Costs-type	Improve	Cash	Customer Value(0to10)
Strategy	profitability	focus strategy		
	Deployment 50%	30% lower sales (product not compliant to the strategy)		
		30%x1mljx10$x40% (margin) = 1.2mlj	1200K$	0
Product Turnover	performance	speed		
		4mth to 3,5mth→ 1mljx10$ → margin +12.5%		
		1mljx10$x40%=4mlj → +12.5%	500K$	3
Know How 50%	30% Idea	Idea Creation		
	Ideation/concept 50%	High margin-Low margin		
		30%x1mljx10$x(40%-20%) = 0.6mlj	600K$	5
Assignment	Quality 75%	pro-activity		
Creation OEM 75%	FCR +1%	pro-activity		
Realization OEM	Quality 75%	pro-activity		
		FCR(+1%) 1mljx1%x10$ = 0.1mlj	100K$	10
Introduction	TTM delay	pro-activity		
		or Brand Index decrease		
	Support	Customer Care		
		4weeks: 1mlj/(4/52)x10$x40% = 0.3mlj	300K$	5
		Brand index decrease	?????K$	10

Explanation:

Strategy deployment is bad (50%, only the highest management level))

-Lower sales 30%: loss of 40% margin: 30%x1mljx10$x40%=   1200K$

Product turnover

-Turnover faster from 4 to 3.5 months. The profit on the
same money is 3 times(3x1000K$) or 3.5 times(3.5x1000K$)= delta =   500K$

Idea creation

-High margin(40%) – Low margin(20%) and 30% is new idea creation.
30% x 1mlj x 10$ x (40% - 20%) =   600K$.

Creation

-Creation only 75% good resulting in additional FCR of 1%
1% x 1mlj x 10$ =   100K$

Introduction

-TTM delay margin loss during 4 weeks
4WK/52WK x 1mlj x 10$ x 40% =   300K$

Total   2700K$

Costs-profit originated by T&D / R&D

T&D / R&D: Why use pro-active DFSS?

Facts:	Costs-type	Improve	Cash	Customer Value(0to10)
NP Creation	Retries TTM	R&D	800K$	
	Design retries	Retest-fees	Knowledge	
			Measurable milestones	
	4WKs TTM loss 1mlj/12x10$x40%		300K$	5
Epidemic FCR	Rework	Test coverage		
FCR>	2% of volume			
	2%x1mljx(10$+10$)x0.5yr		200K$	10
		Devaluation costs RMA		
	2%x1mljx10$x0.5yr/30%(devaluation)		30K$	
Statistical FCR	0.3% to 0.5%	R&D		
	0.2%x1mljx(10$+10$)		40K$	8
DFM (margin)	TS (Repair)	R&D Robust design		
	TS=1000$/20/10=5$/h=1$/TS			
		0.5%x1mljx1$=	5K$	8
	Material loss	0.5%x1mljx10$=	50K$	
	LO increase	0.5%x1mljx10$=	50K$	

Explanation:

NP creation

-TTM delay margin loss during 4 weeks

Real sales loss 4WK/52WKx1mljx10$ (very bad for the company)   **800K$**

4WK/52WK x 1mlj x 10$ x 40% =   **300K$**

Epidemic Field Call Rate

-RMA 2%FCRx1mlj x 0.5yr x (10$ + 10$) =   **200K$**

-or devaluation costs 2%x1mljx10$x0.5yrx30%=   **30K$**

-8Ds 5x8Ds/product/YR ⊠ 5 FTEs = 42days =2mth x 3K$ **6K$**

Statistical Field Call Rate

-FCR 0.2% higher: 0.2%x1mljx(10$+10$)=   **40K$**

DFM with higher product margin

-Trouble shooter 1$/TS x 0,5% x 1mlj=   **5K$**

-Material loss 0.5%x1mljx10$=   **50K$**

-Line output 0.5%x1mljx10$=   **50K$**

Total 1181K$

Costs-profit originated by Factory

Factory: Why use pro-active DFSS?

Facts:	Costs-type	Improve	Cash	Customer Value(0to10)
Lean Mfg	MP time improve 0.5% to 5%	R&D Line output(LO)		
	LO = 6600/10h =	5.5sec/unit		
	LO 0.5% to 5%	1mljx0,5%x10$=	50K$	
		1mljx5%x10$=	500K$	0
Statistical FCR	0.3% to 0.5%	R&D		
	0.2%x1mljx(10$+10$)=		40K$	8
Material	5% to 10% material more expensive	R&D		
	5%x1mljx10$=		500K$	0
FPY	99% 0.5%	Factory R&D		
	TS=1000$/20/10=5$/h=1$/TS			
	0.5%x1mljx1$=		5K$	8
	Material win	0.5%x1mljx10$=	50K$	
	LO increase	0.5%x1mljx10$=	50K$	

Explanation:

Lean Mfg
- **Line output 0.5% to 5% higher**
  1mlj x 5% x10$=**   **500K$

Statistical Field Call Rate
- **FCR 0.2% higher: 0.2%x1mljx(10$+10$)=**   **40K$**
- **8Ds/product/YR ⊠ 5 FTEs = 42days =2mth x 3K$**   **6K$**

Material price
- **5%x1mljx10$ =**   **500K$**

First Pass Yield
- **Trouble shooter 1$/TS x 0,5% x 1mlj=**   **5K$**
- **Material loss 0.5%x1mljx10$=**   **50K$**
- **Line output 0.5%x1mljx10$=**   **50K$**

Total 1151K$

Don't start to calculate now to get precise numbers. It would mean that you don't understand the purpose of the potential profit estimations. The numbers as such are not important. They are signs to show you that with everything what we do and everything what we don't do we will pay a price, a huge price if we don't know what we are doing and if we are not the specialists in our domain. Nothing is for free. The relative value is important and the idea of what it implicated as loss or gain when we get for example from factory the first pass yield (FPY) results. The FPY reducing by 0.5% will have a direct profit of 100K$ and an effect on the FCR(field call rate = returned products) of 0.1% meaning 20K$. The total cost for 1 million units is 110K$. If you get this result for 1 shift of 5K units it means 510$ per shift. I suppose you can hire some well trained engineers to avoid these issues, isn't it? Note also that we didn't put the most important value inside. What is the most important value? Customer satisfaction. With all the above issues we violated the customer satisfaction and that price, we will pay much higher than the values we calculated. Once a customer has a negative feeling about a brand it is nearly impossible to restore the confidence.

Are the customers happy with what we deliver?

Do they buy new products after a first purchase from your brand?

A very high quoted business key performance indicator for a company is customer retention. We have to put everything in place to avoid the "none compliance to customer" resulting in the above costs and even more, we have to put a system is place that is capable of sensing very early in all processed where things go wrong. Just the attitude to measure the performance after doing something is not good enough because the capture of a "none conformance" when the "none conformance" occurs, is too late. The "none conformance" happened and damage has to be repaired. This debug scenario at design release or sometimes screening scenario at mass production is always a worst scenario. Can the developer solve the problem 100% or will he screw up something else with the risk that it is not detected during release testing? Most quality managers will say that we have to set up a screening process. For screening it is more than likely that the screening doesn't detect 100% of the problem. The end user gets a portion of it. We were unable to set up a detection system to capture the potential problem.

All depends on what we want to achieve but if we want a certain quality in our work we have to do something more and people need to have the passion to do things the best, not just common, not just every day work. It has to be the best work. It must be done with passion for the end result, pleasing a customer, even if this customer is a next colleague in a chain of processes. I will try to give you a methodology to think differently about occurring problems. When problems occur something was wrong. Don't look at it as everything went right and you were unlucky. You have to tackle the problem in a way that it will never happen again. Something went wrong, you or the organization did a bad job. Do not hide the problem because it will happen again and again. Put all your pride aside, report the problems and use appropriate processes to tackle the problem once and for all. Here lays a major potential solution. Are the processes appropriate? For what I know all processes are re-active. We measure after we executed the process. We don't measure if we had a full image of all the items that have an influence on the inputs of the process. If we would have a 100% view on what has an influence on the inputs and we would monitor these items, we could control these inputs within certain tolerances and as a consequence we would have controlled the tolerance on the process outputs. Our daily life is also a sequence of processes influenced by parameters, controls and inputs. What is the secret of being successful in daily life? You can be very lucky but we all know that luck is not eternal. You have to be smart and control your life. Don't let life control you. A very common example is that we all know that we have to pay taxes and we should know also the amount we have to pay. The employer already retains taxes before paying the net salary. If the employer doesn't deduct enough monthly we will have to pay a large amount on the yearly tax control. For some people it is really painful because they weren't aware about the total tax due per year. How could we avoid the pain of having to pay a large tax amount? We could have calculated the year tax so that we could have saved monthly or we could have paid more monthly. The clue is to do the calculation pro-active so that we know the total amount to pay. Knowledge is the key to less pain. Use your brains, others' brains and be aware about real life.

Companies are controlled by a huge amount of processes and if we don't know what happens or if we don't know the tolerances on the things that control the processes the process outputs will run out of hand and we will get into trouble sooner or later.

Do you know the law of Murphy?

What can go wrong, will go wrong.

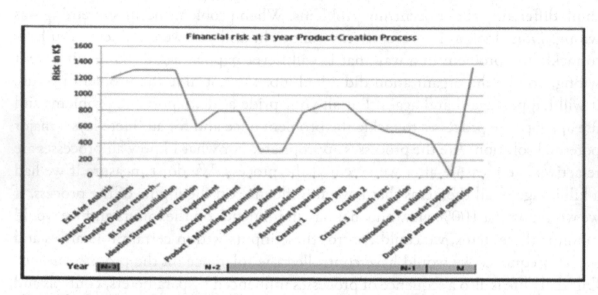

Fig.1 The potential risk of costs or savings at every milestone.

	Cost type	Costs	Costs <--> Customer Value
Business	Strategy	1200	0
	Turnaround	500	3
	Ideation	600	5
	Creation 75%	100	10
	Introduction TTM	300	5
R&D T&D	NP creation	800	5
	Epidemic FCR	200	10
	Statistical FCR	40	8
	DFM(margin)	105	8
Factory	Lean Mfg	500	0
	Statistical FCR	40	8
	Material	500	0
	FPY	105	8

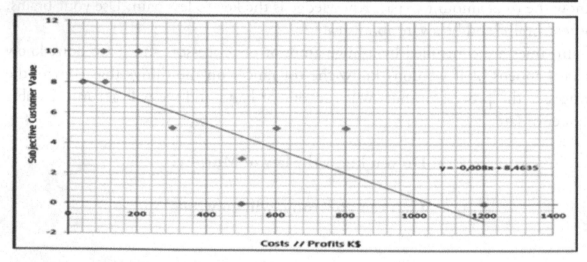

Fig.2 Costs or Profits versus Customer Value.

Overview of the financial risk per milestone over a 3 year product creation cycle, financial risks that can be avoid or eliminated with a pro-active approach to all business processes.

See fig.1 The potential risk of costs or savings at every milestone.

This graph is a graphical representation of the 3 financial "Costs / Profit originated by Business, by T&D or R&D and Factory.

Note: The financial risk for "market support" shows zero because we don't cover marketing nor market support. I am sure there are financial risks at that stage but this is not the purpose of this book.

See fig.2a Highest savings have the lowest customer value.

With "customer value" I want to score the interest for the customer. What is in for the customer if this parameter is improved?

Danger: Highest savings are lowest customer value.
Why care about the customer value?

See fig.2b Costs or Profits versus Customer Value.

When we put all the numbers from the above 3 figures into a sheet and we try to find if there is a certain relation between these costs and customer importance than we can see that there is an undeniable negative relation, see fig.2. When the cost savings or the profit for the company is the highest the influence on the customer is the lowest. If a company is very much involved with creative and new ideas to get high margin the customer doesn't win anything with this directly. The company itself is the most interested partner involved. When the cost savings are the lowest for the company, for example when they create a high quality product with no returned products (low Field Call Rate), it has a very high impact on the customer satisfaction. The customer will be happy with his purchase. Here lays a very high risk. Managers' bonuses are mostly in relation to the revenue performance and thus all the efforts will be put on the high revenue and sales. Many general managers are solely interested in the sales.

Solely focusing on sales profit is digging your own grave.

The most important value is always the customer. You have to find ways to please your customer. If you don't please your customers your sales will slowly stagnate and you will end up in questioning "Why don't we sell?" If we compare our products with the competition there is not much difference. There is a difference. The competition sells, you don't. Perhaps you aren't able to see it. Be aware that everybody will say that they did a good job and that they don't know the difference with the competition.

Just that effect when you purchase a product that you say:

"It works fine and it works even better than I expected."
"It is so easy, this is really what I was hoping for."
"I have it for years already and it is still performing fantastic."

This customer will be loyal and he will purchase products from your brand again. When you invent a new product with high margin this customer will buy your product because he is a promoter of your product.

You have to make sure that all the products are made according to the customer's expectation, not only business expectation, it is customer's expectation because you have to make sure you have loyal customers that buy your high margin products too. Note also that these margin products must have a high quality or the success will be a "one time" lucky story. Make sure that your luck is repeatable and sustainable.

Your good performance must be repeatable.

The company itself has the highest influence on creating loyal customers by creating high standard quality products that comply to the customer's expectations and that comply to the customer's needs. The customer only buys and judges. He can't influence the quality of the product. But the customer's attitude of being loyal to a brand is one of the signposts that I talk about in the "Dangerous Curves Ahead" and that you have to monitor with eagle eyes. I am not going to get deeper into all these 7 signposts but all these curves have one or more relations with customer perception and customer value. Just keep in mind our conclusion from last pages, the customer is the highest value. Many companies have another highest priority, like profit, but know for sure that this other highest priority is wrong. The highest priority has to be "Customer value". This has more to do with management vision and capability. The highest profits will be possible also if we first create a satisfied customer by creating highest quality products.

Stupid example:

Let me give you a stupid example. But sometimes stupidity is better than arrogance with knowledge. Stupidity can be worked around by training and control mechanism but arrogance is a genetic property that is nearly impossible to control.

I have travelled a lot and in the hotels you can see a difference in waitresses' and waiters' customer performance. You can see the employees that like to live and like what they are doing, serving you. Having a hotel with employees that all like to serve the customers, employees with a built in mechanism of kindness and helpfulness, will give you a very good feeling, a feeling like being at home. You will like to return to these hotels. Arrogant critics will say: "The food is the most important." OK. For them the food is the most important but they will also choose for the hotel with kind service with the same quality of food. Yes, food is important but if you have the good food together with the good servants you will win.

When we extrapolate this waitresses example to a product creation process from innovation through development and creation we will all hope to hire the employees with a heart for their work and a heart to surprise the customer with the best product. A company that fails to create these employees is doomed to deliver bad performance.

Pro-activity 2: Epidemic reality
Prevent potential serial killers to kill

When I was writing the book about pro-activity I started reading a scientific article about viruses. The title was: "With an international network that monitors the creation of new viruses rigorously we can perhaps prevent new worldwide epidemics, pandemics."

The word that I was interested in was "PREVENT".

Better to prevent than to cure. At the end of the article there was the following sentence: "Even if we can prevent 1 outbreak of a deathly virus it was worthwhile the whole effort." The costs involved and the amount of casualties we pay to solve a pandemic are tremendously high. If we just think about the SARS and HIV virus and the costs involved. Imagine the whole Asian world with masks, all shops, all buildings with guards equipped with thermal heat sensors to detect if you had a temperature raise.

The place to be is the Central African tribes, South America and Asia. First of all most of our infection diseases are from animal origin. There has to be a carrier to human. In our civilized countries we have less risk to get infected originally by animals but people that have to hunt in the forest are high potentials for diseases. Imagine the hunter who carries his pray over 25 kilometers, sweating heavily with the blood of the pray running over his legs to his bare feet. During his run home he gets small wounds and the blood of his pray enters the blood stream of the hunter. This is how most of the infections happen.

Governments sponsor around 11 projects to monitor the virus population in the tribes and the caught animals. By knowing in a very early stage the new mutations of the virus it is possible to prevent them from getting in the overpopulated world. With the research and the monitoring the researches are setting up prediction models that can warn us for potential pandemics. The whole goal is to detect the serial killer virus before it gets to a pandemic.

Preventing is better than curing.

Pro-activity 3: Nature reality
Acquire knowledge before you act

Clover

Fig.3a Putting elements together

Fig.3b Discovering unwanted results

Fig.3c Analyzing the damage

Fig.3d Solving the problem with a re-active approach

The story about the clover taste honey, clover whiskey and the bees.

I will tell you a short story from nature how pro-activity can mean a prosperous business or fail to harvest and if you want to compete you will end up doing it correctly anyway.

Natural selection by Law of Nature
- **The best learning comes from our MOTHER NATURE**
- **She tells us in a "LIVE or DIE" scenario which is the best way of doing.**

Nature: A certain agricultural area was famous for HONEY and WHISKEY with CLOVER flavor. A farmer tried to copy this success story but didn't succeed in a first approach. For easy reference I call this farmer the "copy farmer".

See fig.3a Clover → Why do certain areas have much clover?

Knowledge: Clovers

> When clovers are young, they are bright and are held upward so that visiting pollinators such as small bees can easily see them and land on them.

Knowledge: Clovers and bees are needed.

> Bees are responsible for the pollination.

> More Bees → More Clover.

> Bees live in Bee Hives.

See fig.3b Populating the bee hives.

Action: The copy farmer buys populated bee hives.
Action: The copy farmer puts the Bee Hives in the fields close to an abandoned shelter because bees are annoying and scaring when they are too close.

Measurement: After a couple of months some Bee Hives are empty?

Less bees → less clover → less and unsuitable honey (no clover honey, no clover whiskey)

See fig.3c The bees left the bee hives.

Measurement: We see mouse droppings. But mice don't eat bees and bees are not scared from mice. This shouldn't be a problem.

Measurement: An unpleasant smell is noticed. The unpleasant urine smell was the trigger for the bees to leave the hive. The unpleasant smell of the mice environment chased the bees away.

How to solve?

Action: The copy farmer buys cats and releases them close to the shelter. Cats chase mice and the problem should be solved.

Measurement: After a while there are no cats left at the abandoned shelter and the problem of the mice stays. I am quite sure that the mice didn't eat the cats

Measurement: Cats will move to the farm as there are mice too and it is much more comfortable there. Why return to the harsh environment in the fields?

How to solve?

Improve: As you really need to get rid of the mice, you need the cat. You will have to move the bee hives to the farm too.

 See fig.3d Solving the problem.

This whole problem is built on re-active behavior. We start thinking when we have a problem.

If the copy farmer would have used DMAIC and studied the product, its environment and its product creation, he would have known the potential problems. Every bee farmer knows that bees and mice are a bad combination. There are even special closures to avoid mice entering the hive.

By doing a study about the bee culture the farmer would have known that his idea to put the bees in the field was a bad idea. Perhaps be using special bee hives he could have solved the problem for a while but know that mice find always the way to food no matter the effort you put to keep them out. The use of cats was a good idea but cats prefer also a comfortable life. By doing a study about cats he would have known also that cats don't stay in the open field if they can get a more comfortable place.

Making the study on forehand would have been pro-active. He could have predicted that the place to put the bees would be critical too.

What is the result for the copy farmer?

He lost Time To Market, TTM.

He lost a year of harvest.

Pro-activity 4: Company reality
Introduce pro-activity in company values

These days, much of what businesses produce and sell can be commoditized. Even the most complex product will find its commoditized benchmarking competitor product.

Customer-centricity can be strategic to win the battle for customers' hearts, minds and wallets by differentiating and gaining a sustainable competitive edge.

But there are chances that you've encountered organizations—perhaps even your own—that don't always seem to care that much about the customer's brand perception. Their phone calls and emails aren't answered. Long waiting time before being serviced. Although this is really the easiest service to please a customer. If your company is still in this phase you are in big trouble. The pro-active approach that I am talking about counts for companies that have already some idea about what six sigma is and they are trying to set up actions to create good products already but they fail for some or for a lot of reasons.

But what about product creation? How much is the product creation team busy with pleasing the customer? Are they solely busy with creating a product that gives them profit or are they busy with creating a product for the customer, knowing that if the customer is happy with their product that the consequence will be that there is profit. This is a complete different approach. In the first approach the team will also release partial good products because they need the profit from this product. In the second approach the team will not release the product because it is not good enough for the customer.

This portion is very difficult. How to design a product that meets the expectations of the customer? Communication is mostly a marketing and sales monologue that leaves little room for the customer's voice to be heard.

But be aware that it is the delivery of the total customer value that drives genuinely loyal customer attitudes and behaviors in a target market, resulting in competitive differentiation and long-term profitable growth for the enterprise.

How can your company avoid inflicting bad experiences on customers? How can you consistently provide great ones? And how can you achieve these goals within your company's existing resource constraints? The answers to these questions are critical to companies that want to succeed in a global marketplace where the competition is always just a click away.

Introduction to good company behavior

Values that are a red wire through the whole book.

Intro 1: The 7 dangerous curves, the 7 dangerous roads to walk on.

    1. Not ready to leap from a "Burning Platform".

    2. Can't link Customer-Centricity to Business results.

    3. Don't know what Target Customers Value.

    4. Wrong Signposts to Keep on Track.

    5. Don't treat Employees like Customers.

    6. Don't use the Whole Company's Brain.

    7. Don't assure that all employees work in the same direction

Fig.I.1. Dangerous curves jeopardizing pro-activity

1. Not ready to leap from a "Burning platform": When things go bad in a certain category you must be able to leave it behind. Take you losses and find a more profitable business. You must be busy with new ideas to take over from old horses. You need high speed horses. Pro-activity means here that you are constantly creating new ideas to take over from old out fazing products.

2. Can't link Customer-Centricity to business results: The management dashboard doesn't have the information between the customer wishes and how this has to be implemented in all the business processes and how it is translated into results. The company results show only dollars but no performance of customer satisfaction.

3. Don't know what target customers' value: This is normally a basic requirement that the Value Proposition House (VPH) should cover. What is the Value that the new product brings? The

Value Proposition House contains the arguments and needed reports to convince management to accept the product for creation. There is a full paragraph about VPH further in the book.

4.Wrong signposts to keep track: If the dashboard only shows financials than this is a typical malaise of management, only caring about the finance. A major asset is to know how well the customer is loyal to the company. What is your customer retention score? Most signposts are re-active. The book guides you to pro-active signposts.

5.Don't treat employees like customers: An employee is the major parameter that makes a process run, that makes the difference between a good company and bad company. Treat the employees like your customers because these employees have to make money for you. A process has suppliers and customers both very important to achieve goals.

6.Don't use the whole company's brain: Huge transformations will be decided by highest management based on financial results but the idea creation and innovation ideas can best be done with the whole company's brain. It is scientifically proven that you need different brain types to come to differentiating ideas.

7.Don't assure that all employees work in the same direction: Even huge companies fail to measure that the teams comply with the business values and customer important values. A team of 10 members with 1 member working in the opposite direction will result in a team performance smaller than 70% because this opposite member will consume the energy of more than 1 person.

Do you have these sensors built in your company and do you have the processes to sense these dangerous curves? Fig.5.1_3 gives a good example of how you can set up a performance measurement.

Intro 2: Pro-activity and scientific prediction are the highest business values

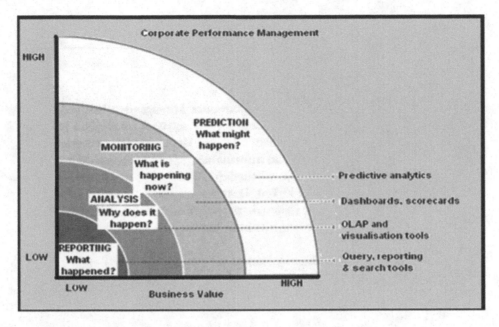

Fig.I.2. Corporate Performance Management highest value → PREDICTION

The highest value in company performance values is mastering reliable PREDICTION. It is a major asset for a company to know in advance which road it has to go and where to put the resources and efforts. Prediction in product creation is like having a correct rolling forecast (rofo) for the business forecast predictions. We will show you with examples and next in the workshops that pro-activity will have a major impact on all these business values, like product quality, strategy compliance and others. We will use statistics to predict failure areas in the product creation process. The prediction about which areas and domains in product creation will give problems is not just prediction. Introducing knowledge, experience, statistics and small experiments in the prediction decision making data increases the probability tremendously. You will be close to 100% sure about the most important problem areas because you will know the relative importance of the problem domains and test areas. With these parameters, knowledge, experience, statistics and experiments, built in we are able to predict very accurate where additional resources should be put, where additional time has to be planned and where a special control has to be build in.

Furthermore we will elaborate more about techniques to measure business process for compliance against the business values and customer values even before the processes are started.

The goal is to introduce a methodology of pro-active thinking that pro-acts immediately on "customer none compliance". The logical consequence is that the 7 dangerous curves will be probably expelled from your company.

Workshop Intro1-2.   Examples of the warnings signs, recognition of dangerous signs.

Chapter 1
Definitions

Pro-activity, prediction and probability.

Pro-activity is a word that we all like very much. It gives us a positive feeling just by using the word. But did you ever think about what it takes to be pro-active? Why would you for God's sake try to be pro-active? Do we know what "Being Pro-active" means?

There are also words quite close the pro-active. Close is perhaps not the best word but having a relation with pro-activity. These words are prediction and probability.

With pro-activity there is probability that the prediction is true but there is also a probability that the predictions fails.

Let's formulate definitions about pro-activity, prediction and probability, just to have a good start on the definition of the names.

1.1: Pro-activity:

Being proactive is about being anticipatory and taking charge of situations.

In Organizational Behavior and Industrial/Organizational Psychology, proactive behavior (or pro-activity) by individuals refers to anticipatory, change-oriented and self-initiated behavior in the work place. Proactive behavior involves acting in advance of a future situation, rather than just reacting. It means taking control and making things happen rather than just adjusting to a situation or waiting for something to happen. Proactive employees generally do not need to be asked to act, nor do they require detailed instructions.

Proactive behavior can be contrasted with other work-related behaviors, such as proficiency, i.e. the fulfillment of predictable requirements of one's job, or adaptive, the successful coping with and support of change initiated by others in the organization. In regard to the latter, whereas adaptive is about responding to change, pro-activity is about initiating change.

Pro-activity is about initiating change.

Pro-activity is not restricted to extra role performance behaviors. Employees can be proactive in their prescribed role (e.g. by changing the way they perform a core task to be more efficient). Likewise, behaviors labeled as Organizational citizenship behavior (OCB) can be carried out proactively or passively. For example, the altruistic OCBs can be proactive in nature e.g. of offering help to co-workers in anticipation, even before they ask, is an example of a proactive action. Other OCBs concerned with the compliance with rules and expectations might even be incompatible with pro-activity.

1.2: Prediction:

A prediction is a statement or claim that a particular event will occur in the future in more certain terms than a forecast. The etymology of this word is Latin (from *præ-* "before" plus *dicere* "to say"). In regards to predicting the future Howard H. Stevenson Says, " Prediction is at least two things: Important and hard." Important, because we have to act, and hard because we have to realize the future we want, and what is the best way to get there.

Outside the rigorous context of science, prediction is often confused with informed guess or opinion.

A prediction of this kind might be valid if the predictor is a knowledgeable person in the field and is employing sound reasoning and accurate data. Large corporations invest heavily in this kind of activity to help focus attention on possible events, risks and business opportunities, using futurists. Such work brings together all available past and current data, as a basis on which to develop reasonable expectations about the future.

1.2.1: Supernatural prediction:

Predictions have often been made, from antiquity until the present, by resorting to paranormal or supernatural means, such as prophecy or by observing omens. Disciplines including water divining, astrology, numerology, and fortune telling, along with many other forms of divination, have been used for centuries or even millennia to predict or attempt to predict the future. So far none of these means of prediction have been proven under controlled conditions and are heavily criticized by scientists and skeptics.

1.2.2: Anticipatory science forecasts:

In a scientific context, a prediction is a rigorous, (often quantitative), statement forecasting what will happen under specific conditions, typically expressed in the form *If A is true, then B will also be true.* The scientific method is built on testing assertions which are logical consequences of scientific theories. This is done through repeatable experiments or observational studies.

A scientific theory whose assertions are not in accordance with observations and evidence will probably be rejected. Theories that make no testable predictions remain proto-sciences until testable predictions become known to the community.

Additionally, if new theories generate many new predictions, they are often highly valued, for they can be quickly and easily confirmed or falsified (see predictive power). In many scientific fields, desirable theories are those which predict a large number of events from relatively few underlying principles.

Quantum physics is an unusual field of science because it enables scientists to make predictions on the basis of probability. Mathematical models and computer models are frequently used to both describe the behavior of something, and predict its future behavior.

In microprocessors, branch prediction permits to avoid pipeline emptying at branch instructions. Engineering is a field that involves predicting failure and avoiding it through component or system redundancy.

Some fields of science are notorious for the difficulty of accurate prediction and forecasting, such as software reliability, natural disasters, pandemics, demography, population dynamics and meteorology.

1.2.3: Finance prediction:

Mathematical models of stock market behavior are also unreliable in predicting future behavior. Consequently, stock investors may anticipate or predict a stock market boom, or fail to anticipate or predict a stock market crash.

Some correlation has been seen between actual stock market movements and prediction data from large groups in surveys and prediction games.

An actuary uses actuarial science to assess and predict future business risk, such that the risk(s) can be mitigated.

For example, in insurance an actuary would use a life table to predict life expectancy. This life expectation over a big population becomes reliable if no parameters change. Therefore it is very important to continue to measure the parameters that have an influence on the prediction to adapt the prediction adequately.

Confidence, Tolerance and Prediction Intervals

When we talk about probability that an event will occur. When we have to get a feeling about meeting requirements just by analyzing certain samples we start talking about confidence intervals. How confident are you that the result from the samples matches the whole lot? When and where are these Intervals Appropriate?

Confidence Intervals:

Typically, since confidence intervals are based upon sample standard deviations, confidence interval calculations require sample sizes of four or more, as recommended by the EPA (EPA/530-R-93-003). Fewer data points result in wider confidence intervals, thus, larger sample sizes are preferred since a narrow interval is more useful. Remember, confidence intervals only apply to parameters, and not individual measurements. Thus, confidence intervals are only useful in estimating what the population parameter, such as the mean, should be; but it does not tell us anything about what any of the individual values in the population range from.

Tolerance Intervals:

Tolerance intervals are more applicable in areas such as compliance monitoring, because they tell us what the individual values should be. If the upper limit of a tolerance interval which is calculated from a sample set is higher than the set standard, then there is a high probability (1-gamma) that more than (alpha) percent of the measurements are above the standard, and thus, that the sight is in violation. As few as three data points can be used to generate a tolerance interval, but the EPA recommends having at least eight points for the interval to have any usefulness (EPA/530-R-93-003).

Prediction Intervals:

As the name suggests, the prediction interval is useful in determining what future values should be, based upon present or past data. Prediction intervals are especially powerful because they can predict what a future compliance point should be less than before it is even collected, as opposed to having to wait until the data is collected in order to determine the tolerance interval and then comparing to standards. Another advantage is that as few as one future sample (k=1) can be used in determining the prediction interval, rather than a sample size of 8 or more for confidence or tolerance intervals. Thus, in areas such as groundwater monitoring, where a long period of time must pass, and few data points can be collected, prediction intervals are especially useful.

A common application of prediction intervals is the regression analysis.

Suppose the data is being modeled by a straight line regression:

$$Y_i = a + b.X_i + C_i$$

where Y_i is the response variable, X_i is the explanatory variable, C_i is a random error term, and a and b are parameters.

Given estimates a' and b' for the parameters, such as from a simple linear regression, the predicted response value Y_d for a given explanatory value X_d is

$$Y'_d = a' + b'.X_d$$

(the point on the regression line), while the actual response would be

$$Y_d = a + b.X_d + C_d$$

In statistics, a prediction interval bears the same relationship to a future observation that a confidence interval bears to an unobservable population parameter. Prediction intervals predict the distribution of individual points, whereas confidence intervals estimate the true population mean or other quantity of interest that cannot be observed.

In other words, an interval estimate of a parameter, such as a population mean is usually called a confidence interval. An interval estimate of a variable is sometimes called a prediction interval.

1.3: Probability

Probability, or chance, is a way of expressing knowledge or belief that an event will occur or has occurred. In mathematics the concept has been given an exact meaning in probability theory, that is used extensively in such areas of study as mathematics, statistics, finance, gambling, science, and philosophy to draw conclusions about the likelihood of potential events and the underlying mechanics of complex systems.

Summary of probabilities	
Event	**Probability**
A	$P(A) \in [0,1]$
not A	$P(A') = 1 - P(A)$
A or B	$P(A \cup B) = P(A) + P(B) - P(A \cap B)$ $\qquad\qquad = P(A) + P(B)$ if A and B are mutually exclusive
A and B	$P(A \cap B) = P(A\mid B)P(B)$ $\qquad\qquad = P(A)P(B)$ if A and B are independent
A given B	$P(A \mid B) = \dfrac{P(A \cap B)}{P(B)}$

Fig.1.3_1 Statistical calculation

This is the shortest theory about probability calculation and I will not go deeper into it because this is not the purpose of this book. The probability for an event to happen depends largely on the number of trials you have. The probability to win the lottery is very small and depends on a lot of factors:

-the number of balls (42 gives a lower probability than 40 balls with the same amount of trials)

-the number of bids

-the amount of bids over the lifetime

If you would live long enough your chances to win the lottery would be close to 100% if you just play at least. If you extrapolate the probability to eternity the probability will be close to a certainty. The problem is that you need an eternal life.

The probability that events will occur can be calculated the easiest with the normal distribution function. By getting sample data you can just create this graphical representation following the probability theory.

In probability theory and statistics, the normal distribution or Gaussian distribution is a continuous probability distribution that describes data that clusters around a mean or average. The graph of the associated probability density function is bell-shaped, with a peak at the mean, and is known as the Gaussian function or bell curve.

The normal distribution can be used to describe, at least approximately, any variable that tends to cluster around the mean. For example, the heights of adult males in the United States are roughly normally distributed, with a mean of about 70 inches. Most men have a height close to the mean, though a small number of outliers have a height significantly above or below the mean. A histogram of male heights will appear similar to a bell curve, with the correspondence becoming closer if more data is used.

For theoretical reasons (such as the central limit theorem), any variable that is the sum of a large number of independent factors is likely to be normally distributed. For this reason, the normal distribution is used throughout statistics, natural science, and social science as a simple model for complex phenomena. For example, the observational error in an experiment is usually assumed to follow a normal distribution, and the propagation of uncertainty is computed using this assumption.

The probability density function for a normal distribution is given by the formula

$$p(x) = \frac{1}{\sigma\sqrt{2\pi}} \exp\left(-\frac{(x-\mu)^2}{2\sigma^2}\right),$$

where μ is the mean, σ is the standard deviation (a measure of the "width" of the bell), and exp denotes the exponential function. For a mean of 0 and a standard deviation of 1, this formula simplifies to

$$p(x) = \frac{1}{\sqrt{2\pi}} e^{-\frac{1}{2}x^2},$$

which is known as the standard normal distribution. When properly scaled and translated, the corresponding cumulative distribution function is known as the error function.

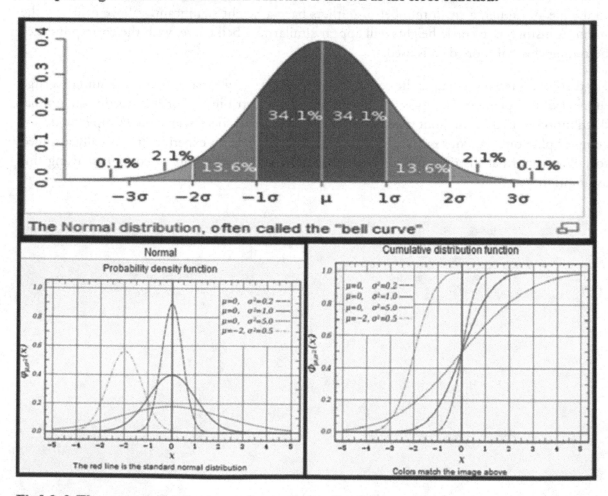

Fig.1.3_2 The normal distribution and cumulative distribution with the six standard deviation (sigma) points around the mean.

The surface between -1 sigma and +1 sigma covers 68.2% of all events measured but you can find more 68.2% surface with other values for example from -2 sigma to somewhere between the mean and +1 sigma. This means actually that the probability that a value lays between -1 sigma and +1 sigma is 68.2% but it means also that the probability that a value lays between –s sigma and somewhere between the mean (mu) and +1 sigma is also 68.2%. The exact values can be calculated with the previous formulas by calculating the integral between of the normal distribution from value 1 to value 2.

Probability in real life:

We can answer the questions easily but note that the answer is always between zero and 1. The answer shall not be 2, 3, ...

Small examples to understand that probability will always be smaller than 1.

Probability worksheet 1

The events can be shown on the probability scale with the least likely being put nearest the 0 and the most likely nearest to 1.

Here are A, F, G and H. Only you know where to put the other letters.

```
    A      F/G                              H
    0 --------1/6----------------------------------1/2-----------------------------------------1
```

	Event	Reason
A	All rivers will dry up next week	This is very unlikely to happen. So it should be close to 0.
B	The sun will shine tomorrow	Your answer will depend on a number of things e.g. the today's weather, the season, and whether you have seen the weather forecast. For example: if it is good weather today you may have decided that it is likely to be sunny tomorrow.
C	You will buy some food today	Only you know the answer to this! However if you are definitely going to buy some food today you should put 1.
D	You will eat some food today	This will almost be certainly be 1, or you could be ill or you could be fasting then you would have 0.
E	You will buy some new shoes soon	If it is likely that you will buy shoes you should put between ½ and 1. If it is unlikely you should put between ½ and 0.
F	Roll a dice and get a 6	The probability is 1/6. If you put your score close to 0 on 1/6, you are right.
G	Roll a dice and get a 2	This is the same as F
H	Roll a dice and get an odd number	The probability is ½. So your answer should be ½.

Fig.1.3_3 Probability smaller than 1.

Probability close to prediction:

Weather forecasters make a prediction of the weather. Why prediction? Because it is in the future. Nobody is sure about what will happen. The next example is extreme thinking and I like that kind of thinking because it shows that you think right. If one would say that he is for 100% sure that the world will still exist tomorrow, he is wrong. There is a very high probability close to 100% (99,999999999999......% and even more) that the world will still exist tomorrow but he can't be sure. I would even be more careful with the global warming threatening the earth. Do you know that the ice age is an instantaneous moment with very high temperature change? If the ice caps melt down and stop the natural warm water flow in the oceans, this could trigger an ice age. So let's write 99.999% probability, less nines.

In the weather forecast you understand the difference between prediction and probability.

Probability Worksheet 2

Watch your local weather forecast each day and notice how accurate it is. For every day that the forecasters get it right reward them with one point. If they get it wrong you get the point.

Correct weather forecast	XX
Incorrect weather forecast	XX

After 10 days who has the most points?

You use these results to work out a probability for the forecast being right.
Probability = (The number of points the forecasters got) / 10

$$\text{Probability} = \text{Correct weather forecast} = \frac{XX}{10} = \frac{xx}{10}$$

The closer your answer is to 1 the better the forecasters are at predicting the weather.

Carry on recording the success of the weather forecasts for another 20 days. Score the weather forecasts as you did before.

Correct weather forecasts for the next 20 days
Incorrect weather forecasts for the next 20 days

You have now checked the weather forecast a total of 30 times altogether. Are the forecasts any more accurate?

$$\text{Probability} = \text{Correct weather forecast} = \frac{XX}{30} = \frac{xx}{30}$$

The closer your answer is to 1, the better the forecasters are at predicting the weather over a long period.

Fig.1.3_4 Weather forecast probability

Workshop 1.3_4 Pro-activity, prediction and probability.

Chapter 2
Why Pro-activity?

2.1: A short Six Sigma introduction and the relation to pro-activity

2.1.1.What Six Sigma stands for:

A comprehensive and flexible system for

achieving, sustaining and maximizing

business success.

Six Sigma is uniquely driven by close understanding of

customer needs,

disciplined use of facts, data,

and statistical analysis,

and diligent attention to managing,

improving and reinventing business processes.

2.1.2.What are the types of business successes?

Cost reduction

Productivity improvement

Market-share growth

Customer retention

Cycle-time reduction

Defect reduction

Culture change

Product/service development

2.1.3. Six essential themes:

1.A genuine focus on the customer

2.Data- and tact-driven management

3.Process focus, management, and improvement

4.Pro-active management

5.Boundaryless collaboration

6.A drive for perfection, and yet a tolerance for failure

2.1.4. Implementation of Six Sigma:

1.Identify core processes and key customers

2.Define customer requirements

3.Measure current performance

4.Prioritize, analyze, and implement improvements

5.Manage processes for Six Sigma performance

2.1.5. Where to Start: Objective, Scope, and Timeframe

Objective 1: Business Transformation

Description: A major shift in how the organization works, aka "culture change."

Examples:

-creating a customer-focused attitude

-building greater flexibility

-abandoning old structures or ways of business, transforming to pro-activity

Objective 2: Strategic Improvement

Description: Targets key strategic or operational weaknesses or opportunities

Examples:

-speeding up product development → note pro-active development in paragraph 5.4

-assuring high quality development → note pro-active development in paragraph 5.4

-enhancing supply chain efficiencies

-building e-commerce capabilities

Objective 3: Problem Solving

Description: Fixes specific areas of high cost, rework or delays

Examples:

-shortening application processing time

-reducing parts shortages

-decreasing volume of past-due receivables

2.1.6. Understanding What Will Qualify as a "Six Sigma" Improvement Project

Why do you need a project?

A major cost or lack of performance makes obvious that a knowledgeable team has to be formed to think about the problem because the problem has a negative impact on the company's performance.

A Six Sigma improvement project startup is needed when the problem has the following properties:

1. There is a gap between current and desired/needed performance.

2. The cause of the problem isn't clearly understood

3. The solution isn't predetermined, nor is the optimal solution apparent.

2.1.7. Six Sigma work methodology: DMAIC

Six Sigma has pro-active management as one of the key themes. Therefore I go further into the methodology that Six Sigma uses and use it as a ground rule for pro-activity.

1. Why-and Why Not-to Adopt the proven Six Sigma methodology "DMAIC"?

Making a fresh start.

Giving a new context to familiar tools.

Creating a consistent approach.

Putting a priority on "Customer" and "Measurement."

**Offering both "Process Improvement" and
"Process Design/Redesign" paths to improvement.**

2. The DMAIC methodology and its content in the 2 types of process engineering:

1. Process Improvement   2. Process Design/ Redesign

1. Define

Identify the problem	Identify specific or broad problems
Define requirements	Define goal/change vision
Set goal	Clarify scope & customer requirements

2. Measure

Validate problem process	Measure performance to requirements
Refine problem/goal	Gather process efficiency data
Measure key steps/inputs	

3.Analyse

Develop causal hypotheses	Identify best practices
Identify "vital few" root causes	Assess process design
	· value/non-value adding
	· bottlenecks/disconnects
	· alternate paths
Validate hypothesis	Refine requirements

4.Improve

Develop ideas to remove root causes

Design new process

· challenge assumptions

· apply creativity

· workflow principles

Test solutions	Implement new process, structures, systems
Standardize solution/measure results	

5.Control

Establish standard measures to maintain performance

Establish measures & reviews to maintain performance

Correct problems as needed	Correct problems as needed

Fig.2.1.7 DMAIC in process improvement and process design or redesign.

2.1.8. Six Sigma project related

1. Creating the project rationale for a Six Sigma project:

A description of the issue or concern

A broad goal or type of results to be achieved

An overview of the value of the effort

Project parameters and expectations

2. Selecting Project "Dos and Don'ts"

DO

Base your Improvement Project selection on solid criteria.

Balance results, feasibility, and organizational impact issues.
Good project selection can be a key to early success

Balance efficiency/cost-cutting with externally-focused, customer focused value projects,
the source of Six Sigma's strength.

Prepare for an effective "handoff" to the improvement team.

Define clear issues and objectives in the projects rationale.

DON'T

Choose too many projects.

Overextend your resources and capabilities.

Create "world hunger" projects.
Better too small projects done more quickly, as long as the results are meaningful.

Put your energies into short-term savings.
It reduces your chance of boosting customer satisfaction and loyalty.

Fail to explain the reasoning for the projects chosen.

Support everything, put your priorities in the rationale.

2.1.9. DMAIC project approach with Pro-active engineering methodology as conclusion

Knowing all the above advantages of the six sigma program we built a methodology to emphasize on PRO-ACTIVITY during the development cycle. You will understand later that our choice for pro-activity was based on the same DMAIC methodology.

DEFINE

We DEFINED the problem

The negative customer appreciation for the products the customer bought.
A major improvement is needed to survive.

MEASURE

We MEASURED the problem

Small pro-active measurement system
versus
a huge costly re-active measurement system

ANALYZE

We ANALYZED the problem

Different teams
worked together
to find the major problems, promoters and detractors.

IMPROVE

We IMPROVED the problem

Improvement by putting pro-active effort
in "Design and all around design"
by problem area and failure area prediction

CONTROL

We CONTROL the problem

The implementation of the new methodology
assures pro-active control
by the measurement system before the process starts.

2.1.10. Major differences between a re-active and a pro-active approach.

Category	Re-active	Pro-active
Measurement	At failure	Deviations
Planning	Uncertain	Very accurate
Maintenance	Corrective	Preventive
Processes	Improve after failure	Acting before doing
Product Creation	Solve at fail	Development quality
Business	Firefighting	Controlling
Marketing	Benchmarking products	Innovative products
Employees	Stressful, firefighting	High job satisfaction
Product	Uncertain quality	Stable best quality

Fig.2.1.10 Comparison re-active ←→ pro-active

The above comparison shows that no-one should be in favor of a re-active business.

Although after reading the words in the pro-active column you will realize that you are perhaps not pro-active at all. Once fallen in a re-active mode you can't imagine how difficult it is to work pro-active. It is even worse because some think that re-active is pro-active and they suppose that they are pro-active although they only measure after events happen. Many systems are set up to measure the performance of prototypes, samples at every milestone and debugging the detected problems. After product release failing products are returned and analysis is done for the next generation. This is a re-active process although many will say that this is the best they can do. To be pro-active you have to foresee the potential problem and act to avoid the potential problem.

Workshop 2.1.10. Your processes → compare re-active versus pro-active with arguments
               Concept to specification process
               Product Creation process
               Product transfer process (between R&D and factory)

2.2: Different added values and their "common sense, real daily life" slogans

These are the ideas that I stand for and for a bizarre reason there are always major constraints within most companies to have these most valuable ideas implemented.

1.Pro-active mindset and methodology in product creation.

Break with firefighting

We all know the word "firefighting". It is trying to solve a problem with extreme priority. Many engineers are pulled from their work and they are pushed to extinguish the fire. In general a 8D is started to make sure it never happens again. An 8D is a tool that gives a good guideline to prevent the problem from occurring in the future. This step is mostly not deep enough. The team should think much further to prevent the problem from happening again. Mostly when this problem is solved the team and the management is happy. Thank God that this problem is solved. Let's wait for the next. The attitude in the 8D should be to examine why this problem happened in the first place and where did our processes or people fail. In most cases we didn't prepare enough or we don't owe the knowledge or experience. Training, covering the lack of knowledge, putting more resources in development are real solutions. But this will not be solved although it should to get out of the firefighting mode.

Preventing 10 times cheaper than curing

When a company has to recall products the costs involved are enormous, transportation, rework, product value depreciation or completely out of the market. It is therefore logical that if we can prevent problems we have to do it. You can't believe how difficult it is to transform processes to pro-active preventive processes. Our natural behavior is re-active. Hope as that we did everything right and if there is a problem we have to be as the best to solve it. In paragraph 5.4 you will see a pro-active product creation process.

2.Pro-active business processes measurement, A0, A1 and A2 processes.

All noses in the same direction

Measurable project and team performance

It is of the utmost importance that all team members or even bigger, all company processes work in the same direction. What do we mean with "the same direction"? It is defending the company values, strategy and core parameters throughout the whole work. One of the most effective ways to achieve this is by measuring the work against these parameters. Fig.5.1_3 shows an example of challenging the processes against the company values.

Transparent reporting, one sheet KPI reporting

Certain units within a company have high interest to get the results from each other. The factory has very important data for the developer. However if this data is not presented in a way that the developer can pull conclusions easily he will not use it. He is developing other products now. Paragraph 6.1.2 gives a good example of useful data representation.

3.Idea creation and Innovation, A1 processes.

Idea creation with the whole company's brain

In many companies the ideas to startup a new product range are born during a general manager's golf session. Or even during a party with top management colleagues from other companies. If these retrieved ideas are subject to the company processes I have no problem with them but in general these ideas are forced through the whole process chain. Ideas have to be generated from specific workshops with knowledgeable people, all type of brains need to be present to get the best ideas. It would give employees a stronger affection to the company if they know that they are an essential part of the company.

Growth by Value Innovation

Value innovation is not the same as idea creation, although an idea could result in a value innovative project. The core of value innovation is to merge properties or products to come to a derivative from the original product but by its specific merged properties give higher value to the customer.

Three values, pro-activity, business processes measurement and idea and value creation will have a major impact on the performance and all on a different level in the total business processes.

Also the sequential order that I use is strange because business processes, an A0 process, should be on top. You will see by reading the book that pro-activity has a major influence on business processes anyway and the word "pro-activity" says more than business processes alone, it means a culture and mindset change. The most important improvement of pro-activity is the attitude and mindset change of the employees and managers.

I will use the "Business process" measurement first to explain the influence of all our added values on the business performance.

2.3: Business Processes measurements

2.3.1. Complete business process A0 and A1 level

We will analyze a complete business process and measure the impact of every core value to the internal processes. In the fig.2.3.1 you see a breakdown of the business processes into A0 and A1 processes. The A0 level processes are the highest processes and the A1 level processes are processes that generate the results for the A0 level process.

Just note that the whole product creation process from strategy to MIP takes around 3 years and the some processes are overlapping in time. This is not a total sequential process. Some processes run in parallel and need the input from others.

Ex. The strategy is needed to create the VPH and the VPH is needed to create the feasibility and to start the product realization process. See fig.5.2_1.

A0 level processes	A1 level processes
BG BL strategy Process	Analyze environment
	Create Strategy Options
	Elaborate Strategic Options
	Consolidate BG Strategy
	Deploy BG Strategy
Planning Process	Investigate White spots & Pain spots
	Generate & Collect Concept ideas
	Enrich Concepts Proposition workbook
	Investigate, screen & define concepts
	Evaluate feasibilities
Product Realization Process	Prepare Assignment VPH
	Concept to specification
	Specifications to drawings
	Drawings to tools
	Tools to product
	Mass production
Market Introduction Process	Product & Marketing programming
	Plan introduction
	Prepare launch
	Execute launch
	Support marketing

Fig.2.3.1 A0 and A1 level process descriptions.

The A0 processes are the highest processes, they are the road to drive your car. If you drive beside the road you will end up with flat tires. On this road several cars can run, the A1 processes and to keep the cars running there are the A2 processes, people, fuel and traffic lights.

2.3.2. Overview of the A0, A1 and A2 processes for a billion dollar company

To get an idea about how a billion dollar company runs, you look at the next sheet with also the A2 level processes. Ex. "Creation of the VPH" and "the concept idea creation" as part of the higher A1 process "ideation strategy" and this one as a part of the higher A0 "Planning Process".

This total A0, A1 and A2 process overview can be put readable on an A3 format paper (the paper format, not the process level). Don't try to read the text in fig.2.3.2 inside, it is not important, just look at the overview of the process levels A0, A1 and A2 and look also at the timing relation between the processes. The processes within the A0 processes are sequential but the processes between the A0 processes can be overlapping.

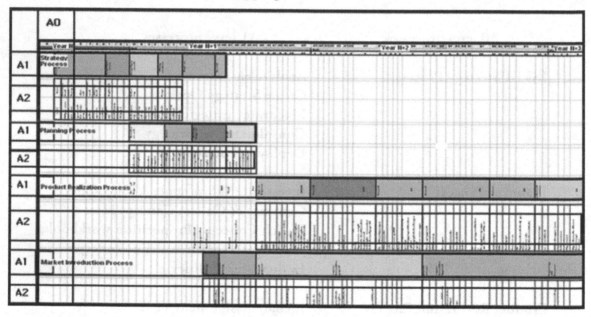

Fig.2.3.2 A0, A1 and A2 level business processes

A0: The highest level processes   A1: The A1 processes are integrated in the A0 process
     BU / BG Strategy process   The processes in the colored blocks
     Planning process
     Product realization process   A2: The A2 processes are integrated in an A1 process.
     Market introduction process   The processes in the black and white cells

Workshop 2.3.2: Create your company's core business processes, the flow

2.3.3. Quantifying the added values

What we will do now is quantifying the impact of our values to the performance of the total business process. How much influence do you add to the process performance by what you do?

What is process performance?

Process performance is the value difference between the inputs and the outputs. You use people, material, equipment and others as input for your process. Your process has a certain performance. You add value with a new approach or an additional input. What is the performance with this change?

What is the "influence"?

A different approach to the processes, like pro-activity, focusing much more on measuring before you start. Next we also use "lean" as a value. Lean sigma is very important for a company as it is based on avoiding all wastes and costs using common sense and knowledge to do things right and thus avoiding waste in product returns and assuring time to market, even reducing time to market.

What are our added values?

How to quantify their performance?

What are the initial values for these "added values"? You can find a relative number for these "added values" as it will be very difficult to put an objective or even a subjective performance value for these "added values" as you can't measure most of them. A better approach is to judge the whole process performance by its normal business result and quantify the improvements on the same processes. We gave a company performance score and predicted the influence of our added values to all the processes in fig.2.3.4-1.

Added Values:

Ideation Innovation

Pro-active Product Creation

Measurable Processes

Lean

Company values: All noses in same direction KPI

Predictive: pro-active KPIs

Fig.2.3.3 The added values for the Company performance score calculation.

See that these values are integrated in the fig.2.3.4 in the top row right side.

Workshop 2.3.3: Create the important values that would impact your company most

Fig.2.3.4_1 Importance per process per added value

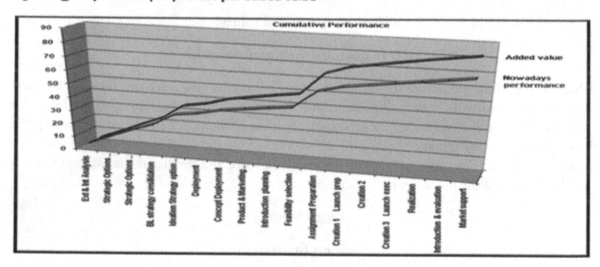

The fig.2.3.4_2 The cumulative performance graph from fig.2.3.4_1

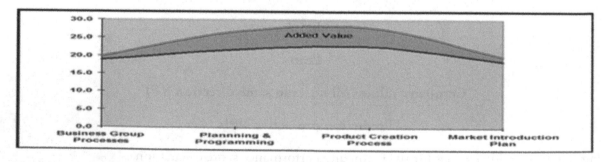

Fig.2.3.4_3 Process performance graph from fig.2.3.4_1

2.3.4. Quantification of the values in the business processes

These "added values" will be used in the following sheet to judge their importance (score from 0 to 10) in all the processes.

See fig.2.3.4_1 Importance per process per added value

The left side of the sheet:

We put numbers of importance for every A0 process, BG-BL strategy, Planning and Programming, Product Creation Process and Market Introduction Process. What is their importance in the total company performance? After in depth consideration we came up with 30, 25, 30 and 15 for the A0 processes (the top row left side 4 main processes) with a total of 100 because the totality of the processes realizes the business performance. The sub-processes are scored on 10 with a totality of 10 for the sum of the A1 processes within an A0 process. The whole idea is that you need numbers to work with. Give it your best shot.

What is the estimated performance of your business? What do we mean with "performance"? Performance is an absolute value of an achievement or fulfillment of a target, like customer satisfaction, like product quality, like customer return rate. How much do you achieve of your parameter? This value can be a subjective own judgment or it can be a well calculated value. Both are Ok if they represent the values that you want to investigate. In our case we are looking to make the customer happy and thus it means "How much did we satisfy our customers?" In the sheet we make this "performance" value variable so that we can verify what the added value influence is at 80%, 90%. The values between the A1 processes with a total of 10 for every A0 process. By adding all A1 process values together we get the process step value. This weight for the process will also be used on the right side but in a ratio 30/70. The weight for the right side numbers is 30 if we compare them with the left side, which is 70. Note however that even if we perform perfectly our added values we can't add the 30%. If we put in all cells the value 10, we will achieve an added value of 30%. To add the whole missing performance with our added values only would be overestimating our performance. You see already that not all cells are filled out and these that are filled out are not 10. There are still other values that are needed to come to 30%.

The right side of the sheet:

As the performance is 70% there is still 30% left to improve. If we do a 100% good job with our improvements we even can't get the 100% performance score because we would need in all cells a score of 10. It would be too optimistic that we solve everything with our added values. There are other values that have to be improved to get a 100%, values that we are not working on. We judge further the improvement for every cell with a score on 10 and relative to the value of the process.

Ex. Assignment preparation Process value is 9*8/10+7*5/10+ ...

(K15*K$4/10+L15*L$4/10+M15*M$4/10+N15*N$4/10+O15*O$4/10+P15*P$4/10).

The result of this sum is relative to a total value of 30 and the row importance, 13, is related to 70.

For the total row we still need to put the row value to it:

"process added value result"=13/70* 17.9 = 3.2

In totality for all the processes we add 12% to the business overall performance with these values. This is a number based on an amount of estimations to come to an overall performance quote. It is not a pure number and doesn't mean too much as number. It should be seen as between 5% and 19%. It is the average value that we would pay for the whole added effort if we would have to hire a company to do it.

Furthermore it can't show the indirect profit from the improvements. But what it shows is the logical relative value that we estimate worthwhile an "added value".

See fig.2.3.4_2 The cumulative performance graph from fig.2.3.4_1

See fig.2.3.4_3 Process performance graph from fig.2.3.4_1

The highest improvement is scored in the middle product creation processes, namely planning and programming and product creation process.

If we do a rough calculation about the influence on added value we can see that nearly half of it is related to pro-activity. What does half mean? It means that nearly 5 to 10% will be improved. The pure number is only the direct related subjective added value. The indirect added value is probably much more important as it is all related to all six sigma values like Cost reduction, Productivity improvement, Market-share growth, Customer retention, Cycle-time reduction, Defect reduction, Culture change, Product/service development and these added values are on the long term much more important. They can be tenfold of the calculated ones.

A best example about the added value is idea creation and innovation. It has a major indirect performance increase. The idea creation processes have in short term a very small impact. However, if we harvest one or 2 successes with new ideas, the impact on the business result will be tremendous if we deliver high quality products. But it would not be logical to start with new products if our nowadays performance sucks. The risk that a new idea will get lost because of our mal performance is realistic, highly probable, even a certainty.

A second example would be the avoidance of epidemical failures. An epidemical problem ended up at the customer's side and passed all the checks. Or not all the checks were done and this was the reason why the problem ended up at the customer's place.

The evidence or the number that we are looking for is more that we are not working in vain. Are we really doing something that normally will turn out to be positive? How positive is the prediction if we use our common sense and most logical thinking? This is the whole question that finds an answer in the above investigation.

Logical Conclusion

PRO-ACTIVITY

Probably the most important value to master

Workshop 2.3.4: Create your added value on your processes
Note: if you can't find any added value you are in trouble. Why would the company keep you?

2.4: Process representation theory SIPOC

S Suppliers	I Inputs	P Process	O Outputs	C Customers
The suppliers who provide the inputs of the process (customers can be suppliers)	The measurable inputs of the process	Start point 3 - 7 high-level process steps Stop point	The measurable outputs of the process	The customers who use the outputs of the process

Fig.2.4_1 SIPOC type 1 Type 1 overview of the SIPOC representation

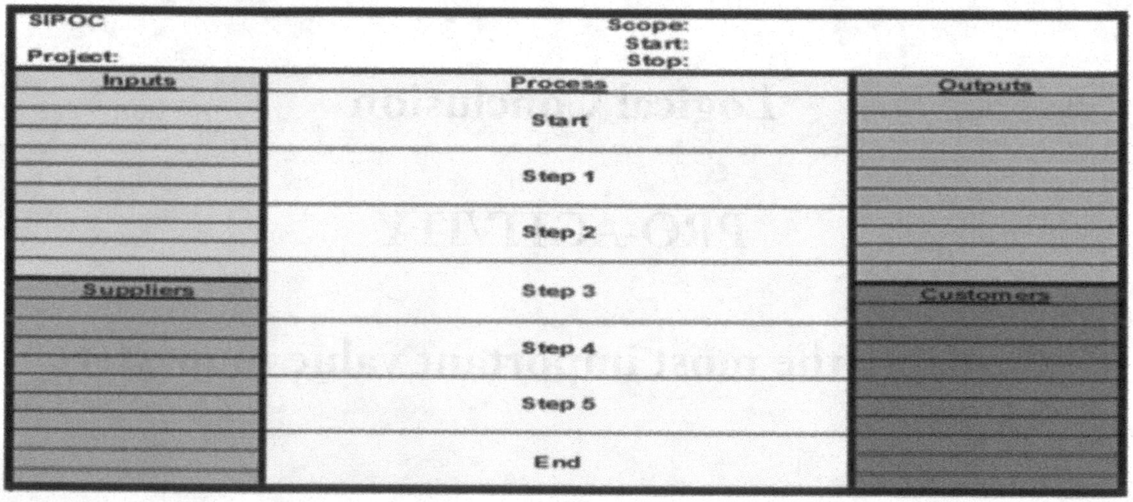

Fig.2.4_2 SIPOC type 2 Type 2 overview of the SIPOC representation.

SIPOC				
Suppliers	Inputs	Process	Outputs	Customers

Fig.2.4_3 SIPOC type 3 Type 3 overview of the SIPOC overview.

We learned already from above that the whole company runs by processes. A reason to give a short theory about process mapping, just a minimum.

The most known representation is SIPOC and it is used to:

-provide a high-level understanding of the process being investigated

-identify the specific start and stop points of the process and prevent scope creep

-identify an initial high-level list of

- **Customers who receive an output from this process**
- **Outputs from the process**
- **Major steps within the process**
- **Inputs that are required to produce the outputs**
- **Those that supply the inputs to the process**

Three different SIPOC representations:

See fig.2.4_1 SIPOC type 1 Type 1 overview of the SIPOC representation

This representation is mostly used as it complies with the basic needs of a SIPOC. Write first the start and stop process and fill out the intermediate processes as clear as possible.

Write all the other reports, inputs, results in the inputs and outputs. Write the inputs and customers to and from the processes.

See fig.2.4_2 SIPOC type 2 Type 2 overview of the SIPOC representation.

I am always in favor to use colors but in this case I prefer this same representation without colors as they don't add anything to the clarity of what we want to show. With this method there is no relation between the process steps and inputs, suppliers, outputs and customers.

Fig.2.4_3 SIPOC type 3 Type 3 overview of the SIPOC overview.

The advantage with this representation is that you can put a relation between the input and the supplier of the input, process, output and customer for the output. This is however not a requirement from a SIPOC but it can help clarity for inexperienced team members.

Example: Call center process

Suppliers	Inputs	Process	Outputs	Customers
Customer	Phone call	**Start**	Call record	Tier2 agent
PMS agent	call	-Customer calls the PMS	in clarify.	FSE
PMS Tier2		call center with a problem.	Service order	Customer
Agent		-Agent answers and creates		
		record of call in clarify.		
		-Agent certifies customer		
		entitlement.		
		-Call transferred to Tier2		
		support. (40% - 69%		
		utilization)		
		-Tier2 attempts to solve.		
		-Tier2 unable to solve.		
		-Agent dispatches FSE		
		to customer site.		
		-FSE goes to customer		
		site and fixes the issue.		
		Finish		

PMS: Problem maintenance service
FSE: Field service engineer
Tier2: second level telephone support

Fig.2.4_4 Example of SIPOC workflow.

Workshop 2.4: Create the SIPOC model for your processes
Note the very strict sequence in the process follow up. The most important step is the beginning of the SIPOC. Define the Start and the End process, the boundaries. Imagine next the sequence of action to get to the End process.

Chapter 3
Cooperation with OEM-ODM suppliers

We talked enough about business processes. Now we will go deeper into the real product creation process and its subcontracting models. One and perhaps the most important influencer to the product quality is the subcontracting model. DO we all "in house" or do we cooperate with subcontractors, suppliers? The pro-active design is most applicable on product creation in a subcontracting model, meaning that your company subcontracts the product design to a supplier in an OEM or ODM subcontracting model.

The problem area is the communication field, the field for misunderstandings, wrong interpretation and supposed agreements. In paragraph 5.4 you will get a detailed methodology to cooperate in a pro-active way with suppliers and minimizing the risk.

The product creation model consists of 1 full In-house model and 3 types of subcontracting

Fig.3.1 No figure needed → All in house, only occasional subcontracting.

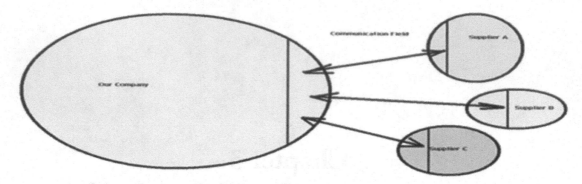

Fig.3.2 Your company manages all is subcontracted.

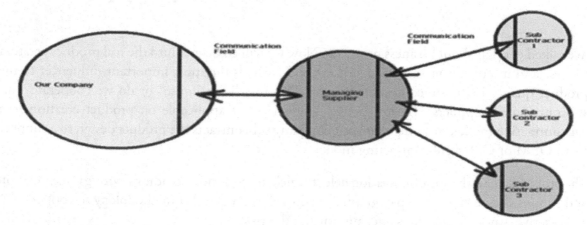

Fig.3.3 One managing supplier

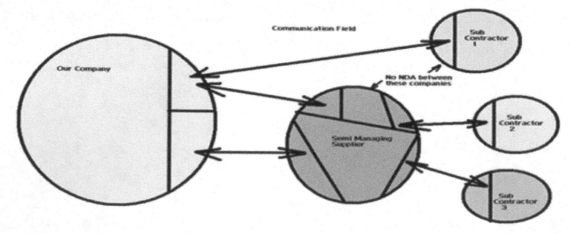

Fig.3.4 One semi managing supplier

3.1: Product Creation with In House engineering, trial runs, testing.

Fig.3.1 No figure needed → All disciplines for Product Creation are "In House" - Only occasional use of subcontracting.

With In-house engineering and testing the product creation quality is per definition a representation of your brand quality. In general it is much more robust because there should be a Design Quality Department that will assure the quality of the design. The company owns the experience and knowledge itself. They will perform the DFMEA better as the full technology knowledge is in house. The risks are known. The problem with this setup often is that it is expensive because the company needs the expensive FTEs in the high cost countries. This cost reason is the major reason why companies will tempt other product creation setups with subcontractors and most of the work executed by the low cost FTEs at the low cost country subcontractors.

3.2: Product Creation with OEM – ODM subcontracting type 1

See fig.3.2 Your company manages all itself but all, design, application, HW, SW and packaging is created by subcontractors.

The product creation is subcontracted to several suppliers and all suppliers deliver their piece of the work. Your company is the gathering point and merging point of all the realizations. The knowledge is actually hired with this approach. The development, trial runs, testing and mass production is subcontracted but the whole product creation process is managed within the own company. This setup is from experience the best approach as you manage your own brand index. This setup saves costs and has the best potential for a certain quality. Many companies suppose that they don't need any knowledge with this setup. Nothing is more true. This setup requires also knowledge. A simple explanation is that you put your company's brand quality to the supplier's quality level as you suppose that the supplier will create the product according to all your wishes and brand requirements.

With this approach it is also obvious that we are looking for low salary countries, in the Eastern Europe, China, Vietnam or already South Africa. This setup introduces many risks because we have to create acceptance criteria for all the work done by subcontractors. But because the managing part is owned by your own company this is somehow the best choice of all bad choices. Many risks still have to be managed with the acceptance tests.

What is the design quality from this supplier? → you need acceptance criteria.

What is the test QA quality from the supplier? → you need acceptance criteria.

What will the supplier do if we force a tight time schedule?

There are minimal requirements for a successful product creation process with this set up. Your company needs the knowledge to challenge the suppliers. If you don't have the knowledge to challenge the suppliers you will get the product with a representation of the brand index of your suppliers.

3.3: Product Creation with OEM – ODM subcontracting type 2

See fig.3.3 One managing supplier

The product creation is subcontracted to a one main supplier and this supplier is leading the whole communication with the subcontractors. The total knowledge is actually hired with this approach. This setup is used to save costs. Therefore it is also obvious that we are looking for low salary countries, in the Eastern Europe, China, Vietnam or already South Africa. This setup introduces a lot of risks that have to be managed although you suppose you only have to take care about 1 supplier.

All "taken for granted" requirements, knowledge, manufacturing capabilities have to be measured and challenged. The same problem as type 1, you need acceptance criteria for everything. Note also that there are 2 communication fields. Although you should not be involved in the communication field with the second tier subcontractors, the communication problem increases as you can be sure that your company's values will not be transposed to these second tier subcontractors.

What are the potential problem areas?

-What is the design quality from this supplier?
-What is the test QA quality of the supplier?
-What will suppliers do if we force a tight time schedule?
-What are the manufacturing capabilities?
      From experience we know the following:
      -The supplier will try to meet the schedule forgetting the quality.
      -The supplier will meet the target due date no matter what happens.
      -The supplier's complying with the customer's questions is a certainty.

3.4: Product Creation with OEM – ODM subcontracting type 3

See fig.3.4 One semi managing supplier

Ex. There is no NDA(None Disclosure Agreement) between a subcontractor and the ODM managing supplier. Because this subcontractor can't do business with the semi managing subcontractor you have to manage the communication between both.

A chipset supplier creates the driver for the windows OS. If we want a special effect, special protection or unique feature we will have to add this functionality to the driver. Software companies create the special functionality and the chipset vendor merges the functionality into the standard driver.
This setup, subcontracting in several layers, is very hard to manage and requires special attention and management, especially because there are always problems to reproduce problems seen on different computers, with different drivers, with different testers, with different operating systems. When compatibility problems arise suppliers aren't capable of reproducing the problem because they are trying to ignore the problem or they don't have the correct equipment. Also test companies only deliver the test performance they are hired for. They aren't hired for fault

analysis. The problem of reproducing the issues falls on your own shoulders. You aren't equipped to test or you don't have the engineers to perform the test.

The communication field is a major problem with this setup. Try to avoid this type of supplier cooperation setup.

Workshop 3.1-4. Create the subcontracting type setup for your product creation process.

3.5: Difficulties related to subcontracting

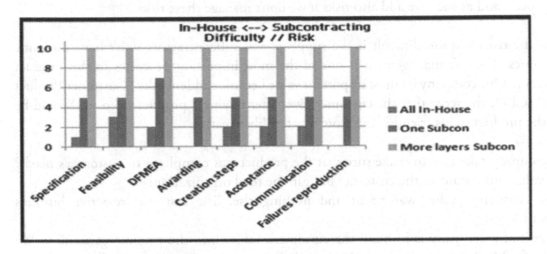

Fig.3.5_1 Difficulty // Risk Comparison between In-house product creation and OEM-ODM subcontracting

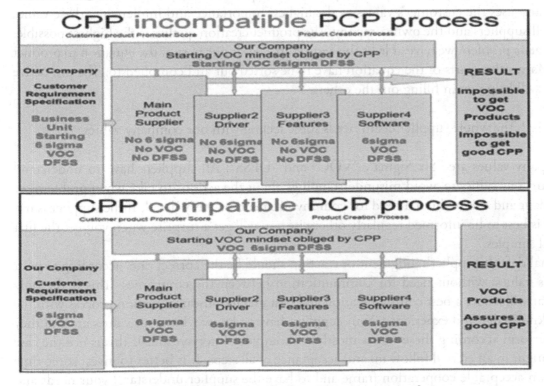

Fig.3.5_2 The whole supplier chain needs some feeling with our company values

See fig.3.5_1 Difficulty // Risk Comparison between In-house product creation and OEM-ODM subcontracting

When we create products with a complete in house engineering team, software engineering, development team and quality assurance test team we have a "built in" system of knowledge upgrade on all levels, a lessons learned process a need to invent and being a leader in innovative products.

If we subcontract in an OEM or ODM subcontracting model the product creation we lose quite some knowledge or we have to hire this knowledge to make sure that the subcontractor – supplier will comply with our brand index quality. We add difficulties in the whole product creation process and as such we add also risks if we don't manage these risks.

Managing the risks is quite difficult if the suppliers and subcontractors don't have the same business values. The risk management is one of the most important processes in this product creation set up. Our company its most important value became suddenly the "Customer Product Promoter" (CPP), the score that the customer gives after buying a product from us. Would he promote the product to his friends? The score was terribly negative.
Why?
Did our company take care to make sure that the product was compliant to customer's needs? Were the values important to the customer built in the total organization?
No. The company's value was profit and nothing else. The rest was re-active business management.
The first task of a company is to build the customer's values into the whole organization. We compare in fig.3.5_2 two types of organizations. Left an organization that partially complies to customer needs. It is impossible to create products complaint to customer's needs. At the right the company uses the customer's values as the highest importance and implements these values between all suppliers and the own company. The product creation quality is the highest possible. If we still get a problem we have at least the best setup. This is one part of the equation to product quality. Many other parts of the equation have to be sorted out and complied to. The pro-active approach will help you in filling out the others.

See fig.3.5_2 The whole supplier chain needs some feeling with our company values

The company values are "Six Sigma", "VOC" and "DFSS". All suppliers have to understand and support our values to avoid misunderstandings about the actions to take when problems or failures occur and this without a need for your involvement for every detail. If the supplier is not reliable it is best to be informed regularly otherwise you will get a surprise when you get the first functional samples.
The normal default supplier's judgment or reaction should be the correct one, according to your company's values, without need for communication between the companies. This is the only system that will assure a best communication and best quality product creation process with the available knowledge and experience. Your suppliers have to know what your values are and they have to perform according these values mostly without your intervention. If this is not the case you are caught in an eternal follow up and acceptance of all steps. It is better to invest some time to set up an acceptable cooperation frame and to have the supplier understand your needs and wishes.

In the next paragraph we will analyze a problem situation between the supplier and your company and we will show a best approach to set up a mutual agreed project as outcome of workshops and mutual cooperation.

3.6: Pro-active process to cope with supplier problems

The supplier construction is as follow: See fig.3.3 One managing supplier with subcontracting type2 and specific problems with solution scenario.

In the cooperation with the supplier we encounter major problems and we consider these as a threat to the cooperation as they jeopardize our brand image. We keep a team workshop and come up with the following problem list as the most annoying in our product creation process during several product creation cycles.

Problem descriptions:

1.Sample release process fails
2.Hardware changes result in failure
3.Function fails after full OEM-supplier testing
4.Competitive pricing
5.Why do we see sometimes epidemic failures?

These problems are supplier problems or cooperation problems. Several 8Ds should be started to tackle the specific problems. However it is not the goal to solve one problem and next time run into another similar problem. We need a structural approach. We have to find the major problem areas or the major areas that will have an overall effect on these problem areas. We need a best and lasting approach. We need structural improvements. These structural improvements have to be solid and measurable so that a stronger system starts.

The cooperative "workshops – trainings" to create lasting improvement projects.

3.6.1.Process charting with SIPOC representation.: Training

It is mandatory to get a good overview of the processes involved in your problem area. Without this good overview of the involved processes the workshop is useless because there is no solid base to talk about. I prefer the type 1 SIPOC model from fig.2.4_1 because there is no visible relation between the inputs and the processes. This is my personal idea. I can imagine that some people exactly want to see this relation between reports and processes. I organized first workshops to give the company the view about process mapping.

3.6.2. Cause and Effect diagram for all the above questions.

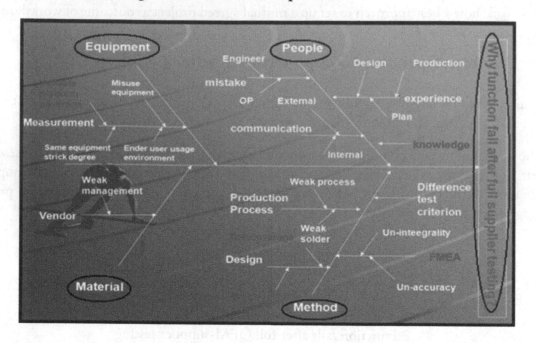

Fig.3.6.2 Cause and effect diagram

The cause and effect diagram shows many problems that have a minor or major effect on the problem statement. Put teams together of 5 people each and give every member 5 stickers to write a cause on the paper. We don't discuss about the value of the idea. After a short silent brainstorm we put the ideas that are related together. Now we write down the remaining causes on the diagram and we judge if the cause has a major or minor influence on the effect. We keep only the major influencers. See the red text.

3.6.3. The summary of the major influencers is the following table.

We form teams to discuss in a cause and effect diagram all the questions that jeopardize the good functioning between the companies. In a brainstorm session we ask everybody to provide 5 problems that have an influence on the problem. After that we summarize only the major influencers. See fig.3.6.3.

The Problems	Summary of major influencers
1.Sample release process fails	Knowledge & Training
2.HW change results in failure	Design mistake
3.Function fails after full OEM testing	HW changes
4.Competitive pricing	Customer requirement changes
5.Why 3 epidemic failures?	Knowledge
	Sample quantity
	DFMEA
	Platform coverage not 100%
	CRS not 100% covered
	Solid design
	Critical mass (economical volume)
	Efficiency
	FPY
	No root cause details & tracking
	Incomplete checks
	Low reliable components
	Equipment & test coverage
	FW bug // design bug
	Training & knowledge
	Many changes
	Incomplete tests // not enough time

Fig.3.6.3 Questions and solutions

When you summarize the major influencers you write all major outcomes from the 5 workshops down and then you try to gather again the common ideas.

Ex. HW changes and Many changes will be put under "Hw/requirements change". At the end we have the fig. 3.6.4 left. We list all the major issues in a QFD diagram to find the solutions that have an impact on more issues and these will have a higher weight in the outcome for the choice of the final project. With this approach we try to get a smallest effort for a biggest result effect.

QFD and project charter

Fig.3.6.4 QFD diagram to get the major influencers for the problems

DFSS				Customer Importance	HOWs (Title) — How to improve the VOC demands													
					Knowledge & training	Designmistake	HW / req change	Sample quantity	DFMEA	TestPlatform coverage	CRS coverage	Solid design	Critical volume	Efficiency	FPY	Revise/case details	Incomplete checks
		Direction of Improvement			○	○	○	○	○	○	○	○	○	○	○	○	○
WHATs (Title)	DFSS improvements		Sample release process fail at C	8.0	◉	◉	◉	◉	◉	◉	○	◉	△	△	△	◉	◉
			HW change result in fail	9.0	◉	◉	◉	◉	○	○	○	◉	△	△	△	◉	◉
			function fail after OEM testing	9.0	◉	△	△	○	◉	◉	◉	◉	△	△	△	○	◉
			Delay of projects	7.0	○	○	◉	△	○	○	○	○	△	△	△	△	○
			Pricing competive	9.0	○	△	△	△	△	△	△	◉	◉	◉	◉	△	△
			Customer had 3 epidemic fails	9.0	◉	△	◉	◉	◉	△	△	◉	△	△	△	◉	◉
			Why corruption (not epidemical)	8.0	◉	△	○	◉	◉	△	△	◉	△	△	△	○	◉
		How Much															
		Organizational Difficulty			9	9	9	9	9	9	9	9	9	9	9	9	9
		Weighted Importance			435.0	295.0	352.0	349.0	417.0	227.0	195.0	485.0	171.0	171.0	171.0	285.0	417.0
		Relative Importance															

Fig.3.6.6 an example of a project charter

1.Business Case:	2.Problem Statement:
The strategy from the board of management is to improve the OEM product release productivity. One strategy is to improve the Product Release Process for the 2 largest OEM customers. Initial financial forecast: Reduce the release process retry and rework loop to ONE resulting a saving of 200K$ and TTM gain. Examples: Design retries, cost of rework, new mold for dimension mistake.	The Product Release Process takes several reworks and retry loops costing 300K$ and TTM delay in average of 4 weeks.
3.Goal Statement:	4.Scope:
Reduce the number of retry and rework loops to zero. Primary metric: CRS(Customer Release Score)(#reworks) -Logistic -R&D release -TTM delay Secondary metric: -Rework costs	Prototype Review till CR

5.Project Plan:

	WKs	Start	End		
Define	2				
Measure	4				
Analysis	4				
Improve	2				
Control	4				

6.Team:

Process owner	RD		
Team member 1	Financial		
Team member 2	Sales		
Team member 3	PM		
Team member 4	Purchasing		
Team member 5	PE		
Team member 6	RD (EE)		
Team member 7	RD (ME)		
Team member 8	RD (SE)		
Team member 9	RD (TE)		
Team member 10	QA		
Coach	Champion / BB		
Sponsor	High Mgt		

3.6.4.Apply the QFD to find the major influencers for the problems

See fig.3.6.4 QFD diagram to get the major influencers for the problems

The QFD diagram puts small, medium and high relations between the "WHATs" (What is the problem → left) and the "HOWs" (How to solve the problems → top row). What it really does, is putting a value of 1, 3 or 9 on the relation. The high relations will be obvious when they are added and so the most important ones will be obvious. There is also a customer importance for the problems. But because all these problems were a reason to start this workshop there is little difference between the importances for these problems. The judgment factor should be the customer's effect.

3.6.5.Conclusion

1.Knowledge & Training	2.HW requirement changes
3.Sample quantity	4.Design FMEA
5.Solid design	6.No rootcause details, only workarounds
7.Incomplete checks	

3.6.6.Create a project charter to measure the improvements

Creating a project charter is not one of the purposes of this book. But read the content of the project charter and try to match as close as possible with it. I admit that this is my preferred way of working. Use an example and match as close as possible.

See fig.3.6.6 an example of a project charter

How is this project charter created? The outcome of the workshop shows a weak performance of R&D in the creation and release to factory process.

We decided to put 1 project to R&D.

By measuring the number of retries in the acceptance loop of the customer, we try to improve and measure the R&D product creation process With this measure we have to improve the design quality otherwise the primary and secondary metrics will never improve.

With this generic project we cover all the problem areas we found and these areas have to be improved to finalize our project.

Workshop 3.6. Summarize 5 to 10 problems you have with the supplier
      Create cause effect diagrams
      Apply QFD // Conclude the results
      Create a project charter to improve the problem area

Chapter 4
Pro-active management of the risks.

The risks can only be managed if we have an idea about the content they represent. By putting as much experience, as much knowledge and as much statistics as possible on the parameters that influence the outputs and results, we will get ourselves the best risk assessment. What influences the outputs the most? The inputs and the controls need a detailed description of the work or product we want to achieve in the future so we can get a best prediction about the potential problem areas and predict the risks around them.

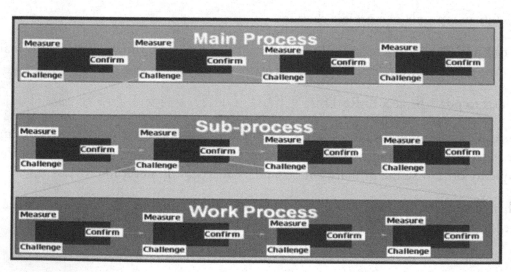

Fig.4 Challenge the input parameters for the process and control the result

4.1: Challenge the delivered work:

We want to achieve a product quality and reliability that reflects our brand image index or we want to increase the brand quality index. What to do? What are the values that are represented in the brand quality? These values are represented in the strategy and in the most valued KPIs used by the company.

If we don't challenge the results we are delivered to the goodwill and professionalism or unprofessionalism of the supplier. How can you be pro-active when working with subcontractors? The subcontractors design the product, PCBA, schematics, plastics, housing and packaging. We have to control the whole planning and assembly of the product by measuring the inputs from the processes. Furthermore we have to control the reliability of the outputs or testresults also.

The following things have to be taken into account:

It is a MUST. You have to:

-take the lead with an OEM-ODM subcontracting setup.
-have the knowledge, you have to know your product.
-have acceptance criteria during the product creation process.
-have a challenging knowledge, testplan and DFMEA workbook.
-know your world otherwise you will be screwed.
-be capable of challenging the delivered work.

When subcontractors are involved we have to be more careful. Subcontractors have their own priorities. These own priorities can be completely out of synchronization with our priorities. Note that our top priority should be "customer satisfaction" and I am sure that this is not the highest priority for the companies from low cost countries. The top priority is there to please you by saying yes to everything. Because saying yes means "complying to get money". Therefore we have to build in our precautions.
We have to make a checkplan, check project quality and this total workbook has to cover the whole product domain, the whole testplans and all the specifications.

But in general we don't have the time, budget nor knowledge to create the above conditions. Without a detailed DFMEA workbook, see paragraph 5.4, you put the product quality in the hands and goodwill of the suppliers. There is always something missing and finally the team decides to see how it will work out. We hope it will work out fine. We hope that the product FCR, the return rate, will be acceptable anyway because not all customers return their products even if they see a problem. Know that hoping is always a bad decision. It is based on failure to control or incompetence. Our customers deserve the best product. If you are not convinced about this you are in the wrong world, the wrong company.

By experience we know that we are unable to guarantee the quality of the products if we are using the OEM-ODM subcontracting model if we don't use special protection mechanisms. Our pro-active product creation methodology will help you a very good step in the good direction by using an adapted tool to co-work with OEM-ODM subcontractors.

4.2:The pro-active challenging effect

We know the definition of pro-activity. We know that we should implement this methodology. But how can pro-activity help to achieve our goals, higher quality, even if we are working with subcontractors?

Where and how implementing pro-activity?

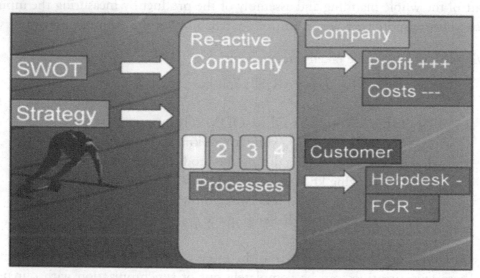

Fig.4.2_1 The most common process execution approach, re-active

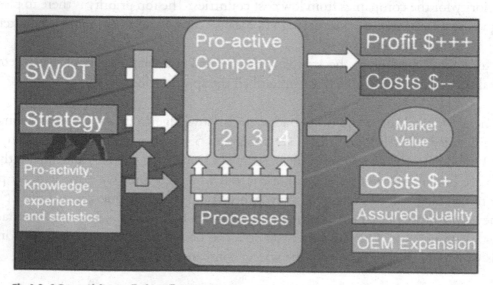

Fig.4.2_2 Pro-activity applied to all processes

1. The company process layout in re-active mode

See fig.4.2_1 The most common process execution approach, re-active

The most common process and business setup is re-active. We analyze the environment, we analyze our competitors, we measure our performance and decide what to do next, this is called SWOT analysis. Why is this a re-active approach? Because the inputs aren't measured on performance. The inputs are the major influencers on the outputs. These inputs must have a quotation against the business values and against performance.

It doesn't mean that all processes can be handled in a pro-active way. See the fig.4.2_2 to understand the difference.

2. The company process layout with pro-activity

See fig.4.2_2 Pro-activity applied to all processes

All the inputs to processes are challenged against a "best off" case, challenged with knowledge, experience and statistics. Did we do our extreme best to make sure that the inputs that we get to steer the process are of the highest quality? I don't mean correct, I mean "Are they as close as possible to the best predictable?" For the predictable portion we need the statistics.

Pro-activity in what?

In all, reports, analyzes, investigations and others where you can add knowledge, experience and statistics.

4.2.1. In measurement:

Input measurements that are directly correlated to the preferred outcome and that they are measurable before the outcome is known.

How can you find these related values that have an indirect influence on the wanted output? Create the process in a SIPOC format and analyze all the outputs and their indirect influencers on the inputs. If there are no obvious influencers, can you create some? Think out of the box. Indirect influencers are also influencers. If a related value changes the input changes too. These are the values we are looking for. Can you make them measureable? In this book you will see enough examples to make items measureable even without exact data.

4.2.2. In planning:

Taking action to cover the predicted and calculated risks that require action in the planning and making that there is enough time to achieve the Time To Market timing or that the timing risks are solved with external help.

We have risk factors that have a direct relation to planning, namely supplier risk and supplier subcontracting difficulties. When the supplier is a block builder without in-depth knowledge of the product, we know in advance that they will just assemble and test after assembly. Furthermore they will not test broad enough. In general they don't know what robustness means. They test functionality and are happy that everything works.

When we know that the suppliers aren't professional enough we have to fill the gap.

Paragraph 5.4 gives a good example of predictive measurement. How much is covered in your testing and how many times will it fail? With these 2 values you can calculate the risk.

4.2.3. In maintenance:

Preventive maintenance.

> By using the lifetime data (DPPM (defect parts per million) and TTF (time to fail)) from equipment and components we can predict the time when the products will fail.

> Also the MTBF test will give a product lifetime indication.

> We can create a maintenance plan for all the equipment and machinery.

Corrective maintenance.

> This type of maintenance is a failure of pro-activity. But let's not be too hard, products fail, no matter how hard we try to get a failure free product. The pro-activity in corrective maintenance could be prediction about the needed support and the needed spare parts.

4.2.4. In product creation:

The most unknown area. Probably you will not be able to imagine what we mean with this but a good example is given in paragraph 5.4. Pro-activity in product creation is the most important portion for a product creating company. It assures the highest quality products.

You will read the explanation in paragraph 5.4.

Workshop 4.2.1-4. The processes in a SIPOC format: indirect influencers (pro-active)
> Measure: which items can be measured and which can be indirectly measured.
> Think about the indirect related measurable values that will have a direct correlation with the wanted effect, the wanted customer judgment.
> Judge all the reports, meetings, measurements against what we want to achieve.
> What do you have to add to achieve the wanted goal? (Ex. Knowledge, experience)

Chapter 5
Pro-active Measurement
&
Probability Prediction

As we anticipate on the events, we act before events happen. What is the advantage of doing so? If we know the events that will happen or have a high probability to happen, we can steer to reduce the probability of these bad things to happen or we can put more effort to avoid the bad things from happening. It is like going back in time to prevent a problem. We influence the quality of the event as we drive the events in the wanted direction with our predictions. The only difficulty that I see now is that we have to find a methodology to get a reliable prediction for the future events.

Ex. If you see someone squeezing his legs tightly you know that in the near future he or she will go to the restroom. And if he doesn't go you can also predict what will happen.

What can be pro-activity here?

When you enter a facility you could try to spot the restroom so that even in a hurry you are not in trouble. Many people locate the restroom when they enter a restaurant. Other people locate the emergency exit in case there is a fire. Someone who knows the exit of a facility has much more change to escape when there is a fire. Pro-activity can be the difference between life and death. Perhaps we should create a checklist before entering a facility. Pro-activity is a built in attitude to avoid potential problems. Perhaps these people are scared of trouble. A valid statement, perhaps pro-active people are scared for problems or for pain. In life situations we should perhaps possess just enough pro-activity to avoid major headaches but in our job we encounter so many problems that cost a fortune, why wouldn't we try to be pro-active in our job?

You have to find elements that give you certain knowledge about the future. Or you have to be better prepared for the future. When a hurricane is predicted, everybody fixes as much as possible their windows, their free moving shelters. We want to avoid trouble in the future by predicting it. In product creation it is really unhealthy to have things happening uncontrolled, isn't it? By having uncontrolled events we have to re-act on these events. We have to correct the trouble caused by going in the wrong direction and we have to put additional effort to go in the wanted direction. The controlled methodology is pro-activity. That is what we are driving for. In pro-activity we measure and act as early as possible. We use additional parameters to build up a case about the future. We take too many risks by doing nothing. We know that issues will pop up. We have to put actions in place to prevent the potential failures.

Pro-active measurement:

How to measure something that we don't have at this moment?

There are several places where pro-activity is most appropriate. The data that we get from some places is so valuable and in many companies it is completely wasted. Let's see what we can do about it. We can retrieve most valuable data from:

1. **The Product Creation Process inputs:**
    Ex. The feasibility study with its values.

2. **Mass production defect data to avoid long time higher Field Call Rate:**
    Ex. 10% of the defects encountered in factory go to the customer.

3. **Indirect but related values quantification:**
    Ex. Market trend index and indirect distributor advice and distributor trend.

We apply pro-activity to several processes or to guideline thinking as:

-all business processes (5.1)
-product creation process: partners and team problems (5.2)
-analyze the problem with all partners, suppliers and the own team (5.3)
-putting workshops conclusions in practice (5.4)
Product failure over lifetime (5.5)
-do not kill a fly with an elephant (5.6)
-the factory is a huge test lab (6.1)
-factory mass production defect control: NFF, NDF (6.1.7)

The above sentences are the guidelines for thinking in the following paragraphs.

5.1: All Business Processes

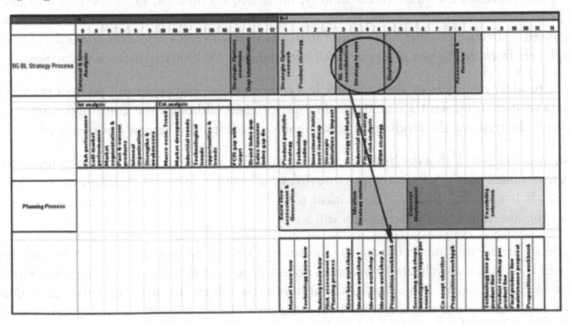

Fig.5.1_1 The 4 A0 business processes with the major A1 sub processes with milestone gates.

Fig.5.1_2 Process time relation: See the yellow cells when (around April) the product manager starts proposing his ideas.

The totality of Business Processes, a 3 year effort, consists of 4 major processes: the BG-BL process, Planning & Programming, Product Creation Process and the Market Introduction Plan. See the summary of the major steps in below figure.

See fig.5.1_1 The A0 business processes with the major A1 sub processes with milestone gates.

An example of an A1 sub process in the A0 Planning & Programming business processes is the A1 Ideation Strategy Options process containing the A2 process "The Proposition workbook revision". The VPH, the result of the proposition workbook revisions, is one of the most important documents to validate in the P&P process. It assures that the product will be according to the business vision, mission, scope and values. In the total business process timing this happens nearly half a year later than the business strategy process has been started. We show in the sheet here below that the BL strategy consolidation and deployment have to be done before the VPH can be created. The product manager needs the strategy input to know which direction the company is going, which type of products he has to create, for which market and which type of customers.

See fig.5.1_2 Process time relation: See the yellow cells when (around April) the product manager starts proposing his ideas.

I would like to challenge the product managers and the business manager to show this validation against the strategy, vision and mission. Many idea creation decisions come from higher management encounters at the golf club. You will say "what is wrong with that?" The whole company processes are violated, no strategy, no vision, no mission, no scope and no values. Ideas have to be created according to the correct process to make sure the company values are respected. Starting a product range out of strategy will absorb more resources and it deviates the employees from the highest company guidelines and their ISO9001 structure in processes.

Some product managers are used to work in a certain pattern and create full product portfolios over and over again with a certain product performance although the company's vision and mission changed and the products to create should be adapted to the new business needs.

A short conclusion is that the business processes should be measured against the most important business values, see fig.5.1_3, to make sure that we all work in the same direction with the same values. It is up to the highest management to agree on the highest values because these value show the direction where the company will go.

There is a logical sequence in the events. The business strategy process steers the whole company. Would it be logical to validate all processes against these strategic values and values like, vision, mission, values, scope? Why would we measure these values? To avoid that the teams don't work focused, to avoid that teams start deviating from our major goal, the strategy and highest corporate values. These are the reason why we are in business. So we have no choice. We have to work in that direction. By measuring compliance to the strategy, vision, mission, scope and values we assure that the whole company works in the same direction.

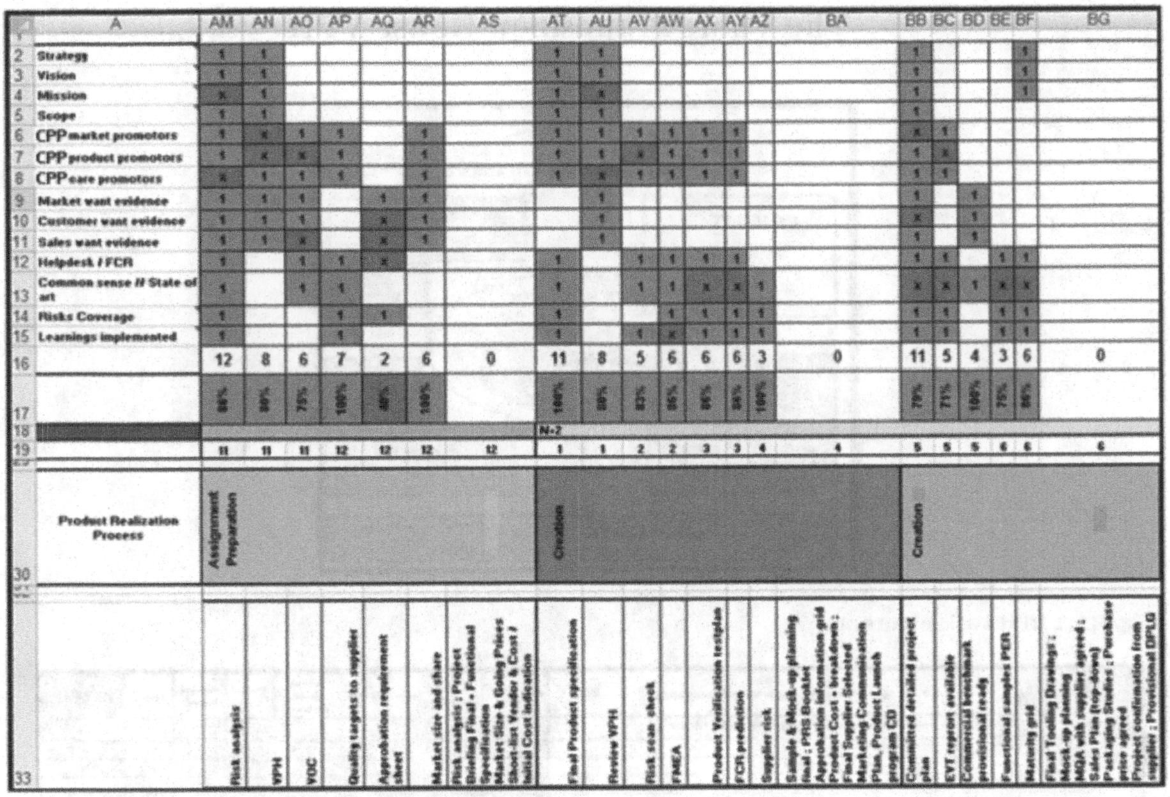

A	AM	AN	AO	AP	AQ	AR	AS	AT	AU	AV	AW	AX	AY	AZ	BA	BB	BC	BD	BE	BF	BG
2 Strategy	1	1						1	1							1				1	
3 Vision	1	1						1	1							1				1	
4 Mission	x	1						1	x							1				1	
5 Scope	1	1						1	1							1					
6 CPP market promotors	1	x	1	1	1		1	1	1	1	1	1				x	1				
7 CPP product promotors	1	x	x	1	1		1	1	1	1	1	1				1	x				
8 CPP care promotors	x	1	1	1			1	1	x	1	1	1	1			1					
9 Market want evidence	1	1			1	1		1								1		1			
10 Customer want evidence	1	1	1		x	1		1								x	1	1			
11 Sales want evidence	1	1	1		x	1		1								1	1				
12 Helpdesk / FCR	1		1	1	1			1	1		1		1			1	1		1	1	
13 Common sense // State of art	1		1	1				1		1	1	x	x	1		x	x	1	x	x	
14 Risks Coverage	1			1	1			1		1	1	1	1			1	1		1	1	
15 Learnings implemented	1			1				1		1	x	1	1			1	1		1	1	
16	12	8	6	7	2	6	0	11	8	5	6	6	6	3	0	11	5	4	3	6	0
17	86%	80%	75%	100%	40%	100%		100%	80%	83%	85%	85%	80%	100%		79%	71%	100%	75%	86%	
19	11	11	11	12	12	12	12	1	1	2	2	3	3	4	4	5	5	5	6	6	6

(Row 18: N-2)

Row 30 — Product Realization Process: Assignment Preparation (AM–AS) | Creation (AT–BA) | Creation (BB–BG)

Row 33 column labels:
- AM: Risk analysis
- AN: VPH
- AO: VOC
- AP: Quality targets to supplier
- AQ: Approbation requirement sheet
- AR: Market size and share
- AS: Risk analysis ; Project Briefing Final = Functional Specification ; Market Size & Going Prices ; Short-list Vendor & Cost / Initial Cost indication
- AT: Final Product specification
- AU: Review VPH
- AV: Risk scan sheet
- AW: FMEA
- AX: Product Verification testplan
- AY: FCR prediction
- AZ: Supplier risk
- BA: Sample & Mock-up planning final ; PRS Booklet ; Approbation Information grid ; Product Cost - breakdown ; Final Supplier Selected ; Marketing Communication Plan, Product Launch program CD
- BB: Committed detailed project plan
- BC: EVT report available
- BD: Commercial benchmark provisional ready
- BE: Functional samples PER
- BF: Mastering grid
- BG: Final Tooling Drawings ; Mock-up planning ; MQA with supplier agreed ; Sales Plan (top-down) ; Packaging Studies ; Purchase price agreed ; Project confirmation from supplier ; Provisional DPLG

Fig.5.1_3 Measuring the process against the key business values

The above fig.5.1_3 shows a small portion of the business processes. The total timeframe to cover the whole process execution is a 3 year overview. The top rows show the business values that we have to think about in every report or evaluation that we do. When we start the process all the involved cells are red and as we progress in the process the reports get completed, we fill out a "1" and the cell becomes green. Note that not all values are relevant for all processes. We target to have an overall compliance of 60% for every column. A score of 60% or higher gives a green result.

Workshop 5.1: Business processes and pro-active value measurement (values from 2.3.3)
 Create the A0 and A1 level business process map.
 Use the excel sheet to summarize all your top (A0) business processes
 Put all the reports, workshop, meeting results as input or output for the sheet

The Value Proposition House VPH

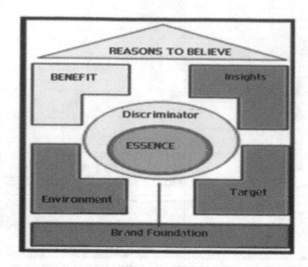

Fig.5.1.1 VPH basic elements

Fig.5.1.2 VPH measured against our business values

5.1.1. The Value Proposition House VPH

Making sure that the products are according to the customer's needs, company's values and according the financial possibilities of the company.

The VPH consists of all the elements that you see in the next picture:

Benefits // reasons to believe // discriminators // insights // environment // target

See fig.5.1.1 VPH basic elements

Benefits: (What are the benefits for buying this product?)
- **Best in class (individual/personalized) wearing comfort & design**
- **Best in class sound quality**
- **Best in class wireless link**
- **Durability**
- **Rewarding listening experience (full performance satisfaction, exceeding expectation**
- **In privacy/undisturbed**

Reasons To Believe: **(Why would we believe that this product is commercial?)**

Digital quality	Ergonomic concepts
Enhancement features	Durability features
Brand & design	Reviews

Discriminator: (What is the discriminator for this product?)
- **Innovative performance enablers that make you want to use longer: best in class wearing comfort/ease of use and quality enablers**

Insight: (What are the marketing, competitor numbers we got to say that this is a commercial product?)
- **Potential consumers have low interest in the product category:**
  - **Low frequency of use (only indoor)**
  - **Main consumer needs: better wearing comfort +**
    - **Outdoor: design, durability // Indoor: easy to use, performance quality**
    - **PC/Gaming: compatible**
  - **High expectation and low trust & awareness on wireless**
- **See competitor as the reference for outdoor**
- **See other competitors as the reference for indoor**
- **See our brand as a mainstream brand**
- **Cons. Need Mini Product: performance quality, portability, ease of use**

Environment: (Where do we position this product, location and price?)
- **Alternative choices**
  - **Listening to alternatives are higher quality**

- • **By pack products**
- • **Products are sold from Euro 2.99 up to Euro 499.99 (and Euro 5.000: Top brand)**
- • **Competitive, mature market. Outdoor: replacement. Indoor: initial purchase.**
- • **Content/source is becoming smaller and smaller with higher capacity and PC based**
- • **Convergence of sources: indoor PC-AV, outdoor: mobile–PDA-camera**
- • **Wireless (BT?) outdoor will come available, from one of the players in the market.**
- • **WiFi will be the dominant indoor wireless connectivity platform**
- • **Power of the trade is increasing, OEM is always their "escape"**

Target: (Who is our target audience?)
- • **Innovators, Plugged ins, Selectives, Classics. Pragmatic approach towards mainstream.**
- • **Everybody who listens to target performance content**
  - • **Home Use enthusiasts // Gamers**
- • **People that want to communicate in privacy or with improved quality /enjoyment**
- • **Main communication target:**
  - • **Outdoor: young innovators & plugged-ins // Indoor: (Young) innovators and Selectives**

Workshop 5.1.1: Create a VPH
> **Use the VPH from 5.1.1**
> **Use the business values as evaluators and score it**

5.1.2. Scoring the VPH

How can we make sure that the VPH is according these business values?

The sheet from fig.2.1.2 shows most of the values for a product. To make sure that the proposed product is worthwhile starting an investment we have to measure its compliance to

The first thing to do to make sure that it is done according these business values is that we will have to measure the VPH if it is designed taking into account certain values.

The second question to ask is: "What do we want to achieve with the VPH?" because this is what we want to happen in the future. The rest we don't care about and thus the rest is less relevant for the pro-active approach.

We put the business values and targets in the top row of the below sheet:

> -Helpdesk and FCR // NPS promoter // Vision // Mission // Customer wants // Market want // Minimal default → these are the business values that should have been taken into account.

> -we have to quantify our compliance to these values. It is normal that not all items will have any relation with these business values. We only need a certain amount, we need that the values have been taken into account.

Ex. Take the VPH "Benefits" → what are the benefits of this product? These benefits have to reflect also our business values but not for the fully 100% because we would never have a product. We can subjectively judge that we need 30% units for the whole "Benefits" block. This is a subjective judgment but it makes sure that the VPH is created considering these values. It is not so much the value that has to be exact, the number has to represent the needed weight of the "Benefits" following the business needs.

We quote the VPH of the product and find out if they match with a certain percentage of the template that we created for the product range. The new product has to comply with minimal compliance criteria. The product has to be judged against all the important business values and give a pass or fail. We put logical value estimators to create measurable conditions. How we set these up depends on the importance and the subjective feeling we judge for every value.

Ex1: Minimal 2 values have to be present in the item.
Ex2: The total amount for the block subject (benefits) should be > 30%

By doing so we know for sure that the team has taken the values into account and they seriously considered the items.

See fig.5.1.2 VPH measured against our business values

The VPH is only 1 example where this measurable approach can be applied or let's say better has to be applied. All processes can be challenged and measured for compliance against the business needs. If you don't do this you will never have a stable company where all teams are working in the same direction. The job will be executed dependent on the employee, architect, product manager and they can't be measured. By measuring you pro-actively measure that processes run as they should run. Is everything put in place to make it happen with best quality?

Workshop 5.1.2: Measuring a VPH
    Use one VPH to measure against values or create a VPH
    Use the business values or specific values as evaluators

5.2: The Product Creation Process

The product creation process is the process best suited for pro-activity because we can best introduce knowledge, experience and statistics in all we have to deliver to the process at the milestones. In contradiction with the generic business processes where we use very subjective estimators to generate Pass / Fail criteria. In the Product creation process we can use experience, knowledge, latest techniques and statistics to validate the reports, decisions and inputs as these reports, decisions and inputs require architectural knowledge, experience and statistical knowledge and this knowledge can be transposed on prediction data. We have many external parameters that change the prediction to pro-active work. The product creation process is a sequential flow of events with the only condition to fulfill the reports and to pass the milestones sequentially. The content of the reports has to be of the highest quality covering all standards and test criteria.

5.2.1. Product Creation Process Re-active versus Pro-active:

The below sheet shows a representation of how the milestone steps mostly are, re-active , and how they should be to convert to a pro-active methodology.
Note the columns with the "Re-active(R) → Pro-active (P)" that indicate the "NOWADAYS" method of working and the "SHOULD BE" or preferred pro-active way of working.

		Product Creation Process	NOW Re-active → pro-active(P)	SHOULD BE Re-active → pro-active(P)	Short reason why.
N+2	Feasibility	Proposition workbook	R	P	Needs evidence for sales, from marketing
		Value proposition house	R	P	Evidence for customer need
		Voice of the customer	R	P	Marketing analysis
		Feasibility scan	R	P	Forecast evidence, distribution commitment
		Approbation requirement sheet	R	P	Countries and sales volume
	Assignment Preparation	Risk analysis	R	P	Supplier, MOQ, critical components
		VPH	R	P	Evidence needed
		VOC	R	P	Evidence needed
		Quality targets to supplier	R	P	Actions to achieve results. Needs measurement for completeness.
		Approbation requirement sheet	R	P	Countries and sales volume
		Market size and share	R	P	Evidence, distribution commitment
		Risk analysis ; Project Briefing Final + Functional Specification Market Size & Going Prices Short-list Vendor & Cost / Initial Cost indication			
	Creation	Final Product specification	R	P	Needs total coverage evidence
		Review VPH	R	P	Evidence needed
		Risk scan check	R	P	Supplier, MOQ, critical components
		FMEA	R	P	Needs to be complete
		Product Verification testplan	R	R	
		FCR prediction	R	P	Needs evidence about target area
		Supplier risk	R	P	Measure the risk
		Sample & Mock-up planning final ; Booklet Approbation information grid Product Cost + breakdown ; Final Supplier Selected Marketing Communication Plan, Product Launch program			

Fig.5.2.1_1 Re-active versus pro-active in processes 1

Read at the right side the reason why there can be a pro-active approach or what we should focus on to get the approach pro-active. The steps should all be validated against all business values before starting.

	Creation	Committed detailed project plan	R	P	No plan based on pro-activity
		EVT report available	R	P	Need also prototype tests, needs acceptance
		Commercial benchmark provisional ready	R	P	How differentiate?
		Functional samples PER	R	R	
		Maturity grid	R	P	No detailed challenging workbook
		Final Tooling Drawings ; Mock-up planning MQA with supplier agreed ; Sales Plan (top-down) Packaging Studies ; Purchase price agreed Project confirmation from supplier ; Provisional DPLG			
	Creation	Pilot run samples approved	R	P	Needs detailed spec before pilot run
		FMEA follow up	R	P	As not detailed value is low
		Service plan	R	R	
		DVT report available	R	P	Total product testplan could be tested earlier
		Quality checklist at DR			
		Field user test	R	R	
		Pilot run samples ; Final packaging & IFU Pilot run samples approved ; Field / user tests Ramp-up planning to supplier implemented ; Confirmed shipments			
	Realization	Post CR action plan	R	P	We can't afford CR failures
			R	P	For EMI pro-active steps can be taken
		Approbation Files			
		Packaging test	R	R	
		Final consumer test	R	R	
	CR	Commercial Samples incl Packaging & IFU Approbation done ; Trial Run (100 pcs) Release Final Consumer Test Digital Product Launch Guide			
	Introduction & evaluation	SPOT, Sales plan on target	R	P	With pro-active approach the TTM is high
		early market feedback	R	R	
		ROFO and startup performance	R	R	
		Filled in evaluation form	R	R	
		Shipment release form	R	R	
		Shipment Release Form, First Quantities Evaluation report Evaluation	R	R	

Fig.5.2.1_2 Re-active versus pro-active in processes 2

Not all processes have the potential to be pro-active. The "early market feedback" can only be done after the products are sold. We can do our best to get the products as soon as possible returned but we can't get the results pro-active.

It takes around 1 year to complete the product creation process. Note that I write that the product creation process takes around 1 year. The total process from idea creation to the end of the market introduction process takes from 2 to 3 years. During this time reports have to be filed, meetings have to be done, measurements have to be completed. The sequential execution of all event results is a RE-ACTIVE process execution. Reports are only validated after execution and the problems are only seen after execution. We execute and measure after execution. The measurement is after process execution, per definition this is re-active.

A lot can be done to overcome the re-activity with personal initiative. All reports, meetings and measurements can be measured against pro-activity values. The processes have to be measured on the inputs. We can also challenge major decisions against the business values (see previous chapter). Mostly only reports are asked, no numbers, no performance against certain criteria. The real challenge is to put these pro-activity values into real numbers. We can measure if all has been done to make sure that the reports will contain what we expect and if these reports will be compliant to validated test report requirements. Did we think about these values when we started process to create the input specifications?

The most important property has to be "Did we do everything we could to know what our product is about before starting the execution of the product creation?" To be pro-active you have to know your product. If you don't know the product you can't challenge your supplier if he performed a good job.

Here is the portion where knowledge, experience and statistics will play a role in pro-activity.

Workshop 5.2.1: Business Processes Re-active versus Pro-active
      Create your business processes overview-flow
      Classify Re-active → Pro-active with arguments

5.3: Analyze the problem with all partners and suppliers.

The company where I worked as Quality manager, a billion company with all possible electronic products you could imagine, got at a certain moment from the highest management the instruction to measure the customer's satisfaction. All business unit managers, all product managers would get a bonus related to this score. Millions were spent to set up a measurement system by calling the customer's impression after buying the product, by calling the customer after repairing the product, by email through web investigation. Let me tell you that it was the same company with a business unit completely not interested in customer satisfaction. The idea was marvelous and well needed by the company because if this idea wasn't invented by the highest management the company would have been bleeding to death slowly. The customer's satisfaction score from the whole campaign was extremely bad for all products, minus 20% in average. Note that I knew that the result would be extremely bad because I was the quality manager for certain products and I wouldn't buy the products myself. The team and the management were focused on TTM (Time To Market) and released the product to market no matter what the quality would be. If the failures would obviously result in returned products the product would be postponed but really this was not often the case. Note that there are issues that can be hidden and because of none competence they weren't able to sense the problems. I was a quality manager with very deep architectural knowledge, so I could feel potential problems, I could also give different test approaches to sense potential problems. But this was always too much. I was called a "Quality Island". I did what I could by using my knowledge to set up traps to catch early the failures but getting involved as early as possible and setting up own testplans. However the whole product creation setup was not adapted to change. The management was a profit management without customer importance and there were some strong personalities, architect and product manager, that made things worse. They worked their whole life already like this. Why would we change? They had always worked like this,

Now with the new KPI, called CPP (Customer Product Promotor) the whole business had to change to the best quality programs, like AQP and DFSS.

I set up workshops with the suppliers and with the team to analyze where they all saw pain points already a long time before this CPP thing was started. The workshop title was: "How to avoid issues at Design Release?"
I will show you the result from the 3 workshops and the summary to come to a generic idea to work on.

The questions to answer were:

Why do we see:

1.failures that have explicit written specifications in the CRS?

2.failures that were implied but not explicit written in the CRS?

3.specifications violations at DR?

4.HW failures at DR?

5.SW failures at DR?

6.architectural failures?

The questions were submitted to teams and the "Cause and Effect" diagram was used to find the influencers that have the biggest impact on the effects.

Workshop 5.3: Product problems from helpdesk and own Quick Feedback Loop
                        Gather all helpdesk calls and analyze
                        Test 100 QRL products or get the failures from at least 100 products
                        Create at least 5 critical questions for a Cause and Effect workshop

5.3.1.The cause and effect diagram:

The team brainstorms about all issues and they put the ideas in a category like People, Equipment, Material, and Method. These 4 categories are the standard categories for a cause and effect diagram. Note that you can decide to create your own categories. The best method is that everybody gets 5 stickers and that you put the stickers on the corresponding category after a 10 minutes thinking period.

The right side shows the question we ask, the problem we investigate: "Why function fails after full supplier testing?" This is the problem that we will try to work on. The supplier did a full release testing and when the products are tested at the customer's side they see problems.

Note the question we asked were the 6 question from the previous page.

All team members give their inputs and these inputs are put under the correct category.

As a team we judge next what the major and minor influencers are. This results in a couple of major ideas, see the red ideas in fig.5.3.1

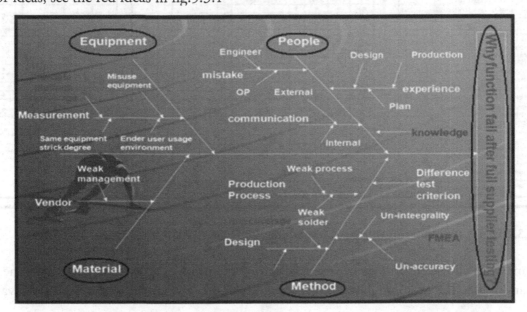

Fig.5.3.1 Cause and effect diagram for 1 problem

After the Cause and effect diagram exercise for all the 6 questions we summarize all the results. See the results from 1 workshop in the top row of fig.5.3.2.

-Training and knowledge   -Test coverage
-PC coverage   -Special drivers
-CRS complete   -and the rest...

For every supplier we create a sheet as below and calculate a relative importance for all issues we encountered during the roadmap 2007 and 2008. See 5.3.3.

5.3.2. Failure summary from helpdesk and returned products:

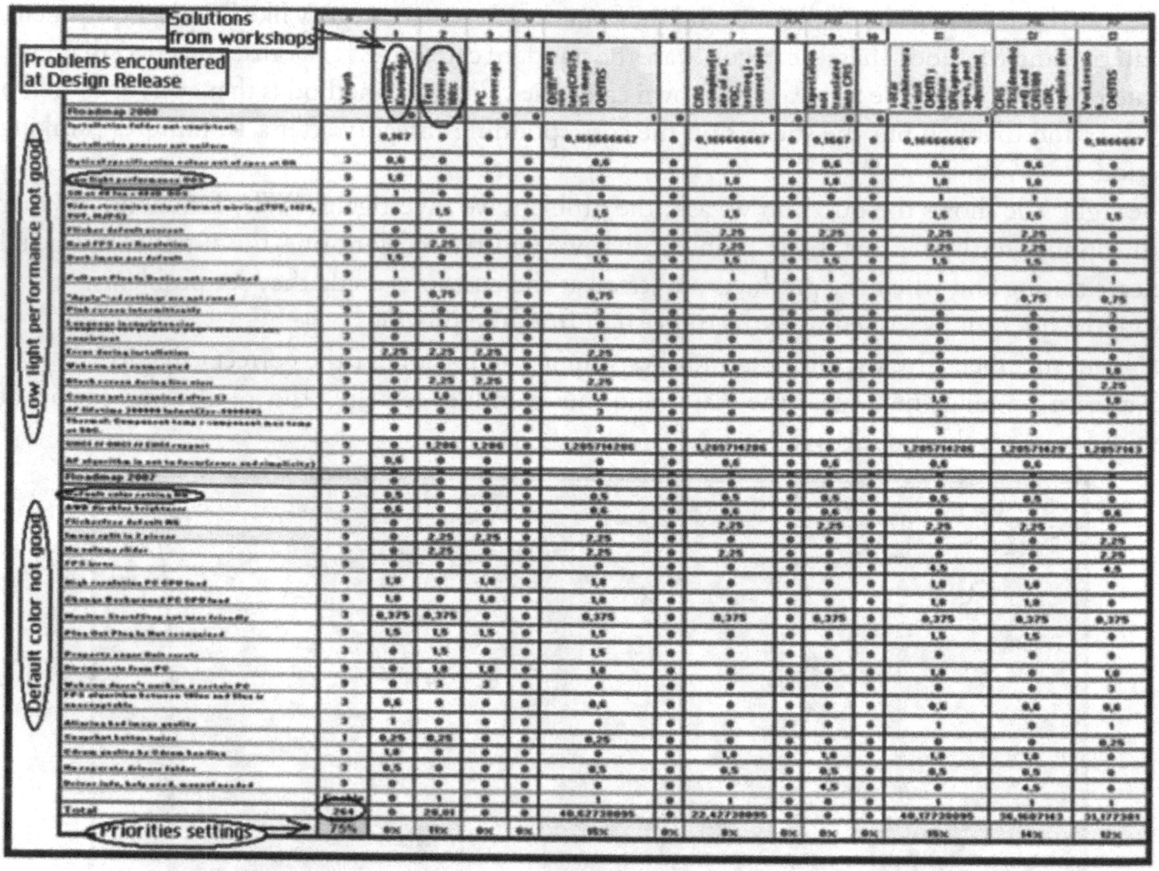

Fig.5.3.2 Issues encountered during the roadmap 2007 and 2008:

A summary of all the failures we experienced during the product creation processes for the year 2007 and 2008. The problem descriptions are not relevant. The layout of the sheet is relevant.

We judge also if the top row solutions have any relation to the solutions that we found as major influencers to our detected failures at design release milestone and during the acceptance testing at the product creation process milestones by putting a "1" in the corresponding cell.

Workshop 5.3.2:   Create the failure summary column at the left.
                   Assign the relation matrix between top row and left column
                   Gather all helpdesk calls, returned products FAs, failure during PCP
                   Create at least 10 issues per roadmap

5.3.3. Assign weights to the problems

Weight: assign a weight value to the failure. Ex. Low light OOS(Out Of Specification) = 9 because it has a direct effect on the user perception of the product.

I only use 1, 3 and 9. This method is used to make a differentiation between the failures.

The weight per "1" from fig.5.3.2 is calculated as follows. On a row with a weight of 9 and 6 "1"s will have for every "1" a value of 9/6=1.5. The more value that a potential top row solution gets the more improvement influence for the solution. The next sheet shows these values.

See fig.5.3.2 Value relation between solution areas and detected issues.

The total value of the weight is 264 (100%). See lower left angle.

For the "Test coverage"-solution we get 11% resolution of all failures we got in 2007 – 2008.

For training and knowledge we get 9% resolution.

Workshop 5.3.3: Weight assignment

5.3.4. Priority assignment

See also fig. 5.3.2 for the explanation of the priority assignment.

We can't put effort in everything. We have to put priorities. We will try to work on as many found solutions as needed to get an overall improvement index of 70%. Therefore we enable and disable some columns by the 1 or 0 in the "Enable" row (lower left side and copied under the potential solution row). We use a logical selection to achieve an improvement of 70% of the problems that we experienced during the product creations of 2007 and 2008.

For this workshop we conclude with the solutions shown in the top row:

1. Test coverage
2. Special drivers
3. CRS complete
4. Total architectural transfer to the supplier
5. Worksession between all suppliers if the product is difficult

In general we would think that "Training and knowledge" will improve the product creation success the most. This exercise is not showing that "training and knowledge" has the highest impact although you can see a score of 9%, a lower score compared to the other solutions. Training and knowledge is a support for a good work but it doesn't create excellent products. The excellent products are created by other conditions if we work in an OEM subcontracting model. We decided that if we do our work better with a detailed customer requirement specification and detailed transfer we had to include training anyway but as a support to create better documents, not as the solution. The conclusion about training is that it will pop up anyway to get this type of workshops going in your company. But suppliers and own employees have to take ownership to improve on the selected areas.

Some big companies start training sessions at the supplier but this is a path to avoid. The knowledge transfer to the subcontractor is dubious when product creation is subcontracted. The risk is real that your subcontractor becomes your competitor and you don't want your competitor to be stronger.

Employees from low cost country companies have the reputation to quit the company when they acquire additional knowledge because their market value increased. So even if the company is loyal and trustful, they have difficulties to keep their employees once they get higher knowledge. The investment will be lost after a year and you will have to train eternally your supplier. A better communication is much more worthwhile doing.

The only system that works is that you put the knowledge in your customer requirement specification and testplans. You can't afford to rely too much on your supplier. If you rely on your supplier you put your brand index to his level. If his level is lower you will put your company's reputation at stake.

Workshop 5.3.4: Create the priority assignment for 1 workshop

5.3.5. Summary of all the workshops

After the workshops with the own team and the suppliers we summarize the results as follows.

Workshop	Category	Issue	%
Supplier 1 workshop	Training	Training, Knowledge	9%
	Test	Test coverage 100%	11%
	Test	PC coverage	9%
	Milestone timing	SW library too late(CRS75%); merge SW/conduct supplier test / I-HOUSE AEAP	15%
	CRS creation	CRS complete(state of art, VOC, testing,) + correct spec	8%
	CRS creation	Expectation not translated into CRS	8%
	Transfer CRS	Total Architectural visit supplier before DR(agree on spec, perf adjustment 100%)	15%
	Transfer CRS	CRS 75%(domeboard) and CRS100 <DR, explicite after to supplier	14%
	Complexity	Worksession SW-supplier I-testhouse	12%
Supplier 2 workshop	Training	Training, Knowledge	7%
	Test	Test coverage 100%	8%
	Test	PC coverage	6%
	Robustness	Different equipment, test method	6%
	Milestone timing	CD: milestone too short to test, measure some spec items	12%
	CRS creation	Material, components can't meet specifications	6%
	CRS creation	CRS complete(install, default, VOC, testing, design Q)	8%
	CC support	IFU-giftbox complete, end-user critical items	7%
	CRS creation	DFMEA	9%
	Transfer CRS	before CD: Sign-off of CRS requirements+Additional step in PPP(Design Q, Architectural)	11%
	Transfer CRS	Transfer CRS to supplier not explicite 100%	11%
	Complexity	Complexity of PCP	9%
Team workshop	Training	Supplier competence	8%
	Test	Testing coverage(SW interaction)	7%
	Test	PC coverage 95%	6%
	Test	USB full compliancy(USB IF, UHCI, OHCI, EHCI)	5%
	Milestone timing	Build in 1 retry cycle. Build in room for failure.	12%
	Robustness	Four corner testing	7%
	CRS creation	CRS need explicit coverage	8%
	CC support	Good manual support and website support	7%
	Test	Measure all spec at CD	5%
	Complexity	Budget to support the need(flicker, AF)	5%
	CRS creation	Design Architectural Quality improve	12%
	CRS creation	CRS translation to end user expectations	9%
	Complexity	Complexity of supplier interaction	9%

Fig. 5.3.5_1 Intermediate summary sheet of the 3 workshops

Category	Workshop	Issue	oem1	oem2	oem3	Total	Rank
Training Knowledge	1	Supplier competence	8%			8%	
	1,2	Training, Knowledge		7%	9%		
Testing	1,2,3	Testing coverage(SW interaction) 100%	7%	8%	11%	17%	3
	1,2	PC coverage 95%	6%	6%	9%		
	1	USB full compliancy(USB IF, UHCI, OHCI, EHCI)	5%				
Robustness	2	Different equipment, test method		6%		4%	
	1	Four corner testing	7%				
CRS creation process	2	before CD: Sign-off of CRS requirements+Additional step in PPP(Design Q, Architectural)		11%		24%	1
	2,3	CRS complete(install, default, VOC, testing, design Q)		8%	8%		
	1	CRS need explicit coverage	8%				
	1	Design Architectural Quality improve	12%				
	1,3	CRS translation to end user expectations	9%		8%		
	2	DFMEA architectural and Quality and VOC		9%			
Transfer CRS to supplier	2	Transfer CRS to supplier not explicite 100%		11%		17%	2
	1	Measure all spec at CD	5%				
	3	CRS 75%(domeboard) and CRS100 <DR, explicite after to supplier			14%		
	3	Total Architectural visit SUPPLIER before DR(agree on spec, perf adjustment 100%)			15%		
	2	Material, components can't meet specifications		6%			
Milestone timing	2	CD: milestone too short to test, measure some spec items		12%		13%	4
	3	SW library too late(CRS75%); merge conduct OEM test / AEAP			14%		
	1	Build in 1 retry cycle. Build in room for failure.	12%				
CC Support	1	Good manual support and website support	7%			5%	
	2	IFU-giftbox complete, end-user critical items		7%			
Complexity	1,2	Complexity of supplier interaction and of PCP	9%	9%		17%	4
	3	Worksession All SW involved			12%		
	1	Budget to support the need(flicker, AF)	5%				
Improve %		Total proposed improvement by yellow numbers	79%	79%	74%	73%	

Fig. 5.3.5_2 Result workshop "How to avoid issues at Design Release?"

The problem is always: "How do we put a sheet together so that we can calculate easily and so that the results will be obvious?" This requires some experience and patience.

We have 3 results, one from the two suppliers each and one from our team. Every team formulated around 10 specific improvements. If we had to put scores on all the potential solutions we would have to take 26 solutions into account. We had to find broad areas to set up generic improvement that will have an effect on future similar issues too.

See fig. 5.3.5_1 Intermediate summary sheet of the 3 workshops

We are trying to find the generic areas to work on. We don't need the detailed solutions. If you work very specific the overall improvement will also be very specific but it will not have a broad improvement effect on similar issues.

We try to categorize the solutions from the 3 workshops into broader working domains. For example all items that have a relation to the CRS creation process can be put under the same category "CRS creation". If you read well the solutions you will find similarities in these 26 solutions to come to a summarized overview of 7 to 8 categories. For example "Competence", "Knowledge" and "training" can be put under the same category. The best way to get this done is to put all the solutions in 1 column and you put generic words, like "test", like "transfer CRS" etc…, that describe the solution in a second row. After some reviews you will come up with a best summary or second best summary. It doesn't matter if you found the best or second best solution. What counts is that you are thinking intensively about "How to improve the problems?" and it is a joined effort that created the solutions. Your suppliers and you will be in favor of the solution and they will put efforts in place to support it and what is more is that you have a cost saving to support the it.

Again, the goal of the whole effort is that we have to find one or two optimized generic areas to work on. It doesn't have to be very specific, a generic area to work on is the best.

See fig. 5.3.5_2 Result workshop "How to avoid issues at Design Release?"

By categorizing the areas to improve we can come to a major improvement conclusion, major areas to improve, all conclusions that are a result of thinking of 3 teams. The 3 companies are much more favorable to implement these improvements because the improvements are the results of their own ideas.

The 3 areas where we should work on are:

1.Customer Requirement Specification (CRS) creation process:

-CRS must contain all domains, must be completely explicit (no implied requirements) covering the complete knowledge of the product, the interfaces, installation, default settings, VOC, test requirements, architectural, design and quality requirements
-the milestone needs a sign off process including a representative assuring that VOC, technical domain, process capabilities (timing and planning), testing coverage and supplier capabilities are met.

2.CRS transfer to the supplier(s):

The transfer from the CRS and translation to the supplier FRS must be a detailed process and interaction between Philips and suppliers covering planning, timing, real measurements and all the above requirements.
Advisable is to create first CRS 75% and transfer the CRS 75% to the suppliers with the explicit knowledge of the 100% requirements but due to the lack of information unable to specify the 100% requirements. Later the CRS must be 100% signed off by the supplier.

3.Product complexity and supplier setup complexity:

Complexity of the supplier setup or subcontractor setup (more than 1) requires additional product creation time.
More suppliers involved in the product creation process requires 6 to 8 weeks additional time to perform a pre-DR from which date still 6 to 8 weeks are needed to solve the bugs found during this first intermediate milestone pre-DR.
A debug session between the chipset supplier and the SW feature supplier at the testlab facility is advisable to debug the "unable to reproduce" problems.
Complexity of the product creation as such (new technology and difficult interaction process between suppliers requires 6 to 8 weeks additional time to perform a pre-DR from which date still 6 to 8 weeks are needed to solve the bugs found during this first intermediate milestone pre-DR

Note: a decision step to define "difficult process" or "easy process" is advisable.

Workshop 5.3.5:   Create first the intermediate summary sheet
                     Reduce the working areas to a ratio 1/4
                     Come to a conclusion

These workshops

are

a merge of the ideas of

the suppliers and the product creation team.

Workshops must have actions.

Act

to improve in

the proposed areas.

5.4: The 4D-CFMEA: Putting the workshop conclusions into practice

The message that we got from the workshops was "being more detailed" in the CRS and the transfer process to the supplier must be much more detailed. I have known companies where the purchasing manager handed over the CRS to the suppliers. The suppliers take a quick look at the CRS, try to see if it deviates from before and accept the challenge. This is the wrong procedure. We have to pay much more attention to details in the CRS and furthermore we have to put more attention to the handling of difficult projects and complex suppliers' constructions.

We have to introduce knowledge, architectural HW and SW knowledge, experience, testplan knowledge and control over the whole process. The suppliers need more detail about the requirements and they need a detailed transfer of the product requirements. This is the responsibility of the requesting company. It is our responsibility. If the requesting company doesn't have the knowledge or if they don't make sure that the whole product domain is covered they put themselves to the knowledge level and capabilities level of the supplier. The requesting company's brand index will be made by the supplier's brand index.

The best methodology is to create a system where we start from generic knowledge to create an overview of the "work to do" and force ourselves by the methodology to dig out the whole product creation process in detail. Furthermore to get the timing and the time needed for complex matters correct we will introduce the prediction effort of problem areas.

The methodology that I worked out is based on FMEA in 4 dimensions. It includes prediction of the problem areas. This comes very close to the requirement and it assures pretty well that the above guidelines will be fulfilled. It goes to the real rootcause of the problems, the major influencers to solve the mainstream of our problems in the cooperation with the suppliers or subcontractors.

The whole operation is based on the creation of an important and crucial block diagram based on blocks that have a complete function meaning that the blocks are testable, are questionable at different suppliers, are architectural units or clustered units that are testable and questionable for full coverage.

Communication between chips will always have a block in the block diagram.

Take block 7 from the fig.5.4.1_1, this block is a communication highway between two critical components, the sensor chip and the video controller chip. This block has to make sure that both components can support the specifications of the CRS.

The four dimensions of the FMEA: 4D-FMEA
1.Dimension 1: The block diagram
2.Dimension 2: The knowledge areas and knowledge build up domain
3.Dimension 3: The testable and controllable dimension
4.Dimension 4: The prediction dimension

5.4.1. Create a detailed block diagram: Dimension 1.

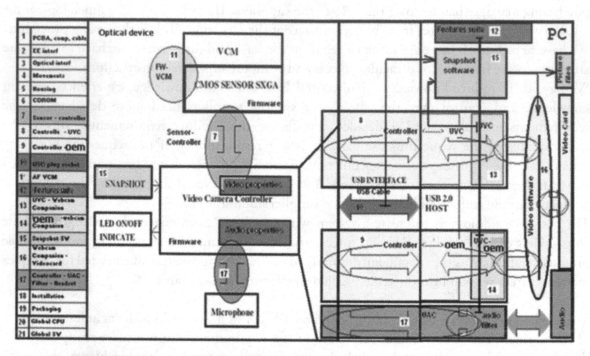

Fig. 5.4.1_1 Block diagram with testable and controllable blocks

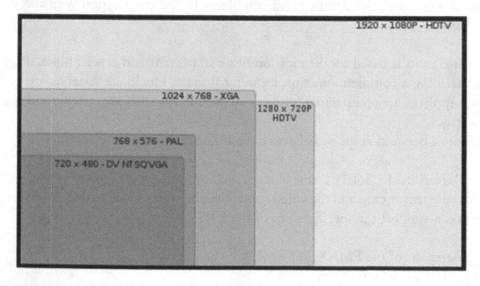

Fig. 5.4.1_2 HDTV specification and other resolutions

See fig. 5.4.1_1 Block diagram with testable and controllable blocks

I decided to break down the total product into measurable blocks meaning that these blocks must have inputs and outputs that are measurable. The inputs and outputs can be SW, HW, bandwidth, throughput or other. The end specifications have to be challenged through these inputs and outputs.

Why?

Ex. See the Sensor and the controller, block 7. It is possible that the end user specifications requires HD 1080P. The sensor can be capable of providing this format but it is possible that the controller can't handle the dataflow bandwidth.

Therefore these blocks have to be made in a knowledgeable manner. The block number 7 "Sensor – controller" needs to provide the information that capabilities requested in the Customer Requirement Specification (CRS) are guaranteed through this block. This block is a protocol, a physical connection on the PCBA and the compatibility between both to assure that the CRS can be met. The sensor can be capable of delivering the format and the throughput but the controller can be unable to handle the flow because he has much more to do. What about this format with higher frame rates? All these questions are explained in block 7.

An example of architectural mistakes that can be waived and the product sold with the HDTV label on the box although the product doesn't comply with the standard.

The CRS requires the product to be compliant the HDTV standard. This means without any compromise 1280 x 720P or 1920 x 1080P.

See fig. 5.4.1_2 HDTV specification and other resolutions

Very late in the product creation process we detected that the controller output was 780P which is not the exact HDTV standard. The product should be rejected. The standard is not met and the controller can't be used. Without deep architectural investigation this product will be sold and is sold.
BTW: the product was sold and it seemed that more vendors sold the product with the same lie. What is the problem when we cheat the customer? Who cares? This attitude will turn against you sooner or later.

Workshop 5.4.1:   Create a block diagram for a product

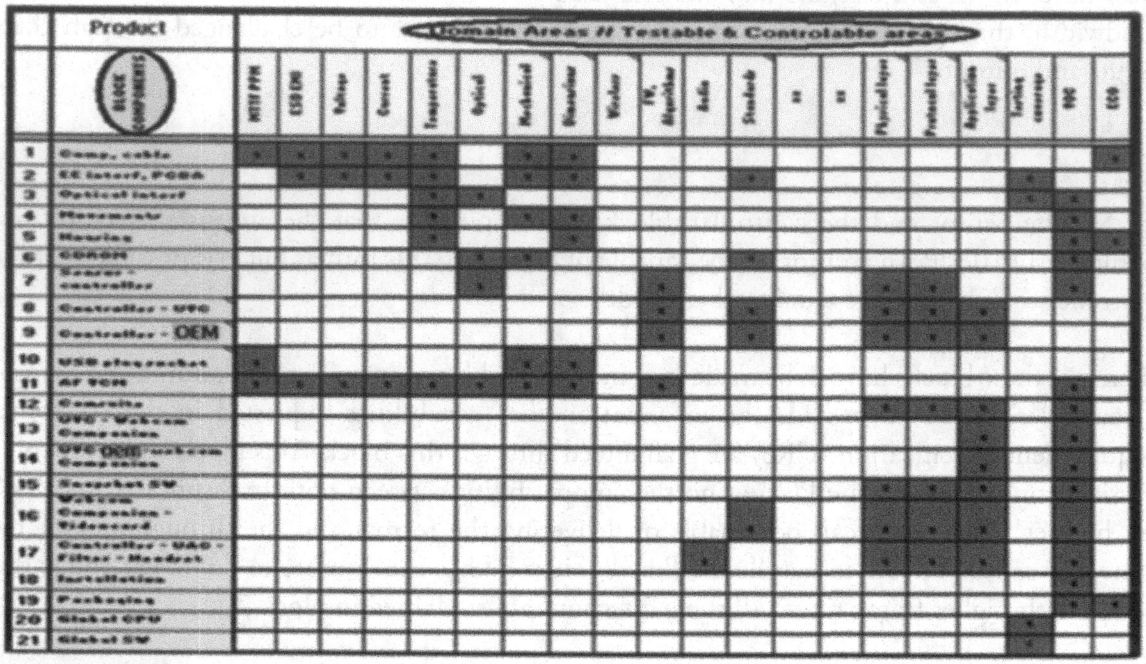

Fig. 5.4.2 Domain – test areas matrix

Fig. 5.4.3 The folders containing the architectural knowledge.

5.4.2. Create the "domain – test area" matrix: The knowledge dimension.

See fig. 5.4.2 Domain – test areas matrix

1.put all the blocks in the first column of the excel sheet.
2.put all the technical field, protocol fields, different other fields in the top row of the excel sheet. These fields are common for all electronic products.
3.put a "1" when there is a relation between the blocks and the test areas.

For every "block components" from the block diagram, an investigation folder is created with a sheet for every domain areas (the "1"s in the sheet), covering the architectural knowledge about the block, covering the test method, the testplan and covering the specification requirements (if possible). The result is the detailed adp-DFMEA workbook.

Workshop 5.4.2:   Create the "Domain – test areas" matrix

5.4.3. Create a folder matrix with all blocks: The knowledge dimension.

See fig. 5.4.3 The folders containing the architectural knowledge.

All folders contain the detailed knowledge, architectural research and guidelines. For every "1" there has to be a very detailed explanation what it really means.
You have to be aware that this is a process that has to assure that you acquire the full and detailed knowledge of the whole product, also the blocks like "global SW". This means that we have to cover all software from the world. How much do you want to cover? 50%, 80%, 100%. What does this mean? How can you make sure that you cover your own specifications?
At this moment we have maximized the architectural knowledge about the total product.
We should also try to find the best testplans too.

Workshop 5.4.3:   Create the folders with all the knowledge

Fig.5.4.4 Lack of coverage sheet

	Product BLOCK COMPONENTS	Lack of coverage / Lack of testing coverage(0% to 100%)													
	Dimension 3	Depth & State of Art	Reference/Margin PxP	Standard(IC RH, ESD, USB IF, WMU)	Performance	Functional	Reliability	Compatibility?	Interoperability	Fixation/Mounting	Software functional	Software robustness	SS	Compliance standards	SS
1	PCBA, comp, cable						5x			20x					
2	EE interf		10x	10x										10x	
3	Optical interf	10x			10x	5x	50x								
4	Movements	1x					1x								
5	Housing		10x				10x								
6	CDROM				25x	1x				25x				50x	
7	Sensor - controller				10x		10x								
8	Controller - UVC				5x	1x		1x	1x						
9	Controller - Oem			5x	25x	5x		25x	25x						
10	USB plug.socket		20x				10x								
11	AF VCM	10x			20x	1x	10x								
12	Camsuite	10x									1x	25x			
13	UVC - Webcam Companion	10x									1x	25x			
14	UVC-Oem-webcam Companion	10x									1x	25x			
15	Snapshot SW	1x				1x	1x				10x	10x			
16	Webcam Companion - Videocard										1x	25x			
17	Controller - UAC - Filter - Headset	10x	10x			1x	1x				25x				
18	Installation	20x										20x			
19	Packaging													5x	
20	Global CPU									25x					
21	Global SW										15x	30x			

Fig.5.4.5 Failure probability sheet

	Product BLOCK COMPONENTS	Failure probability(100% to 0%)													
		Depth & State of Art	Reference/Margin PxP	Standard(IC RH, ESD, USB IF, WMU)	Performance	Functional	Reliability	Compatibility?	Interoperability	Fixation/Mounting	Software functional	Software robustness	SS	Compliance standards	SS
1	PCBA, comp, cable						10x			5x					
2	EE interf		10x	10x										10x	
3	Optical interf	50x			50x	1x	1x								
4	Movements	1x					1x								
5	Housing		1x				1x								
6	CDROM				1x	1x				1x				1x	
7	Sensor - controller				25x		1x								
8	Controller - UVC				50x	10x		1x	1x						
9	Controller - Oem			1x	50x	50x		50x	50x						
10	USB plug.socket		1x				1x								
11	AF VCM	50x			30x	10x	30x								
12	Camsuite	1x									1x	1x			
13	UVC - Webcam Companion	1x									10x	1x			
14	UVC-Oem-webcam Companion	30x									50x	50x			
15	Snapshot SW	20x				1x	1x				15x	1x			
16	Webcam Companion - Videocard										1x	20x			
17	Controller - UAC - Filter - Headset	1x	1x		25x	25x					25x				
18	Installation	25x										25x			
19	Packaging													25x	
20	Global CPU									25x					
21	Global SW										50x	25x			

5.4.4. Create the test domains (Dimension 3) in a new excel top row.

See the "Default & State of Art", "Robustness Margin PnP", etc... These are different items than the top row items from fig.5.4.2. These top row items are all let's say "quality assurance" domains. We score on reliability, functionality but also on very important customer specific items like "default and state of art".

See fig.5.4.4 Lack of coverage sheet

Because we will know the detailed testplans after the architectural investigation we can fill out the missing elements, "Lack of coverage". How much do we not test? What we don't test can fail. The tested elements will detect the problems and don't create a risk.

The quantification is not a precise number and it has not to be precise, it has to show the major problem makers later. Note here also the Low, Medium, High estimation or 5%, 25% and 50% values work also fine. My way of working is as follows. If I can say 1 item, I use 5%, if I find 2 items I take 25% and with more I put 50%.

Workshop 5.4.4:   Create the "Lack of coverage" matrix

5.4.5. Create the "Failure probability" sheet: Dimension 4.

See fig.5.4.5 Failure probability sheet

Why do I put 50% for "Optical interface" and "default state of art"? Because in half of the product releases there is a problem with the optical performance.
Also here this has not to be precise. The relative importance will do its job and show the differences.
If you would use here also only small, medium, high or 5%, 25%, 50% judgments the results would be very similar.

Workshop 5.4.5:   Create the "Failure probability" matrix

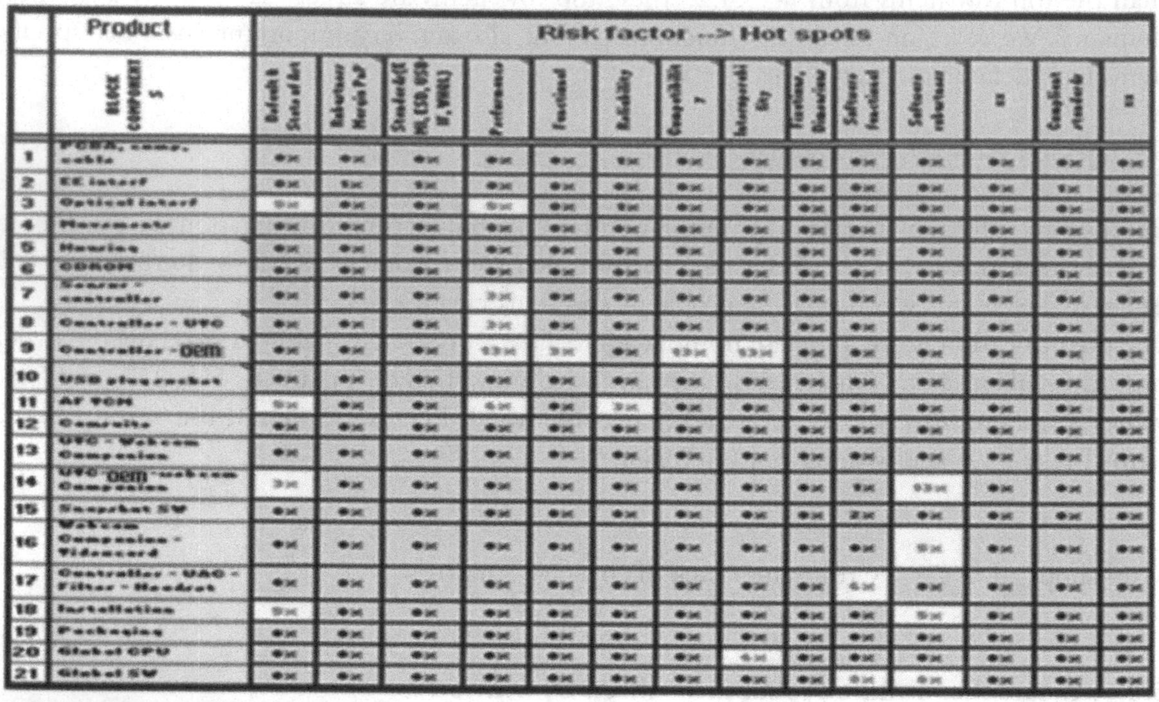

Fig.5.4.6 Risk factors sheet

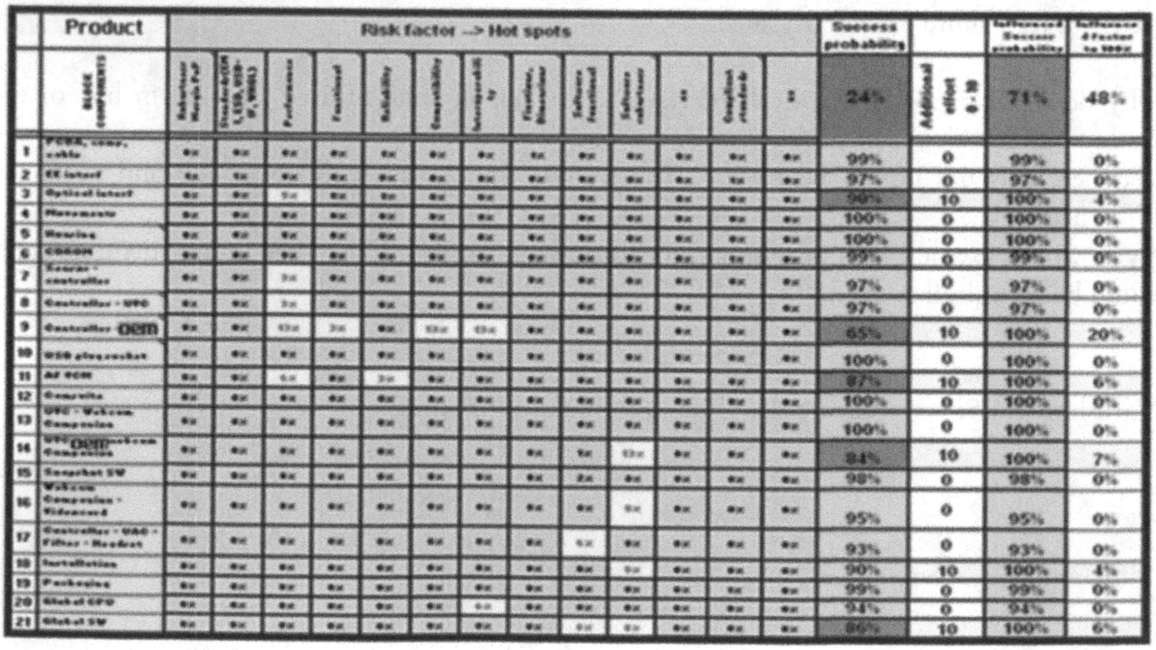

#	Success probability	Additional effort 0 - 10	Influenced Success probability	Influence factor
	24%		71%	48%
1	99%	0	99%	0%
2	97%	0	97%	0%
3	90%	10	100%	4%
4	100%	0	100%	0%
5	100%	0	100%	0%
6	99%	0	99%	0%
7	97%	0	97%	0%
8	97%	0	97%	0%
9	65%	10	100%	20%
10	100%	0	100%	0%
11	87%	10	100%	6%
12	100%	0	100%	0%
13	100%	0	100%	0%
14	84%	10	100%	7%
15	98%	0	98%	0%
16	95%	0	95%	0%
17	93%	0	93%	0%
18	90%	10	100%	4%
19	99%	0	99%	0%
20	94%	0	94%	0%
21	86%	10	100%	6%

Fig.5.4.7 Success probability prediction

5.4.6. Calculate the "Risk factors": Dimension 4.

See fig.5.4.6 Risk factors sheet

The risk factor is "Risk factor= "lack of coverage" x "failure probability""

You can see that some cells become yellow. The trigger value is > 1% but to make the sheet work you have to put the trigger value to a value that only 15 to 20 cells will become yellow. You have to tune the conditional value for the cells so that only 20 cells maximum remain.

Workshop 5.4.6:   Calculate the Risk factors and align the "conditional color" conditions.

5.4.7. Predict the success probability: Dimension 4.

See fig.5.4.7 Success probability prediction

We consider the product creation as a final assembly line in factory where each station(Block) has to do a number of tests(Test areas like robustness margin PnP, Performance, ...) and the result of the product creation process(final assembly) called the FPY(First Pass Yield) will be as follows:

FPY=SP(Success probability total)   = PRODUCT(SPn)   (1<n<21)
SPn(Success probability row n)   = PRODUCT(1-RiskFactor_i)   (1<i<14)
i   = The test areas

With additional effort we can increase this success probability % in the following domains:
1. The optical interface   4%   2. The SW interface   20%
3. AF VCM   6%   4. Interaction SW Companion   7%
5. Installation   4%   6. All Software coverage   6%

We need to get the total success probability higher than 70%(experience value). To get this done we put additional effort (from 0 to 10) on the low scoring blocks. The project leader and the team have to put an action plan with valid actions to assure that the additional effort is met, resulting in a product success probability of 71% for the example here above.
Advantage: the problem areas are known at the project start and the team can assign resources.

Workshop 5.4.7:   Analyze the Success probability prediction for actions

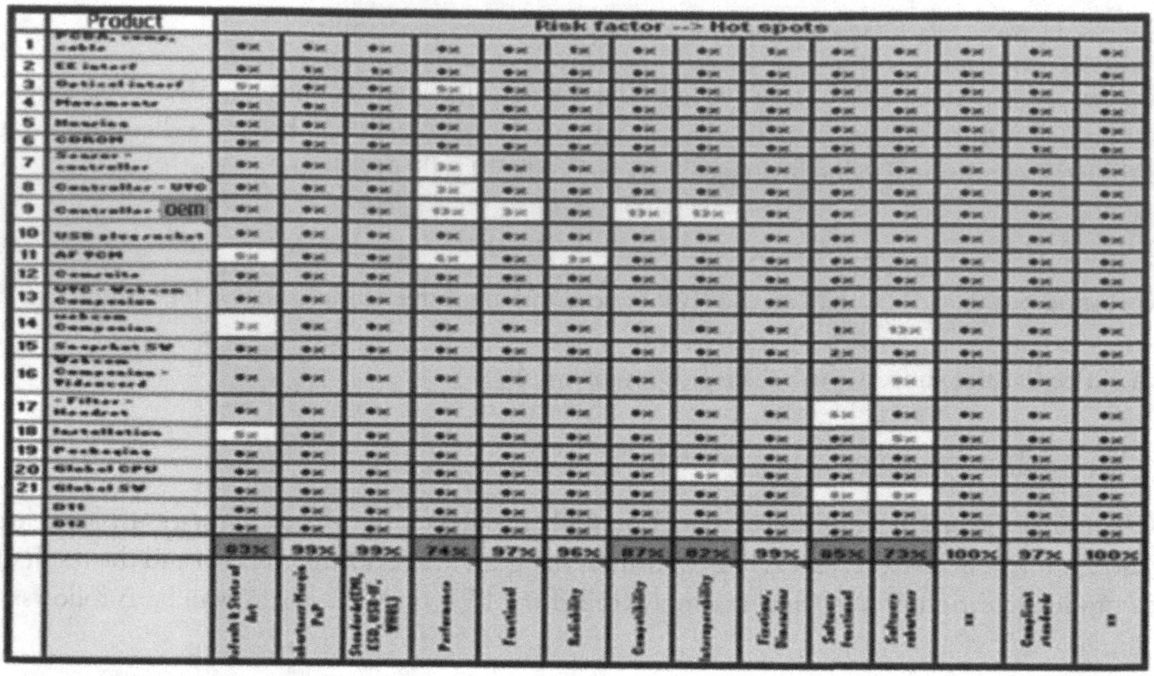

Fig.5.4.8 Problem test areas prediction

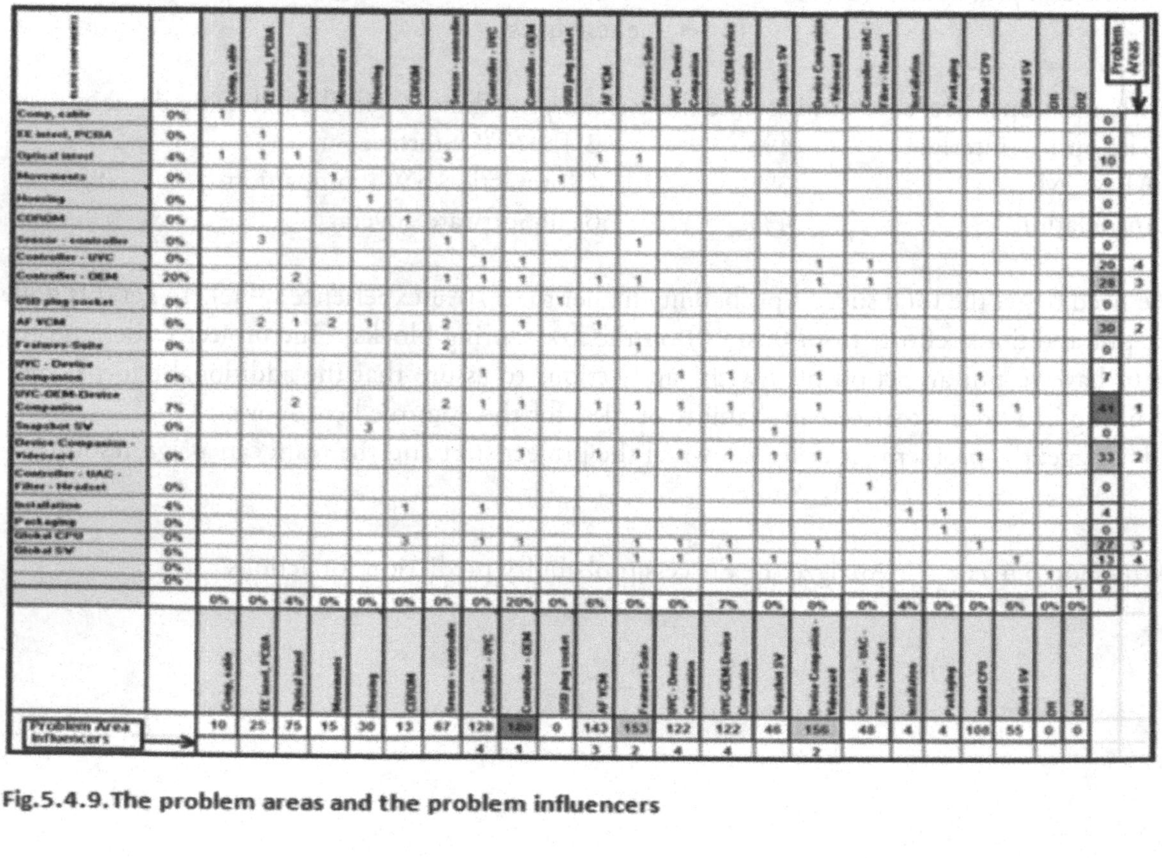

Fig.5.4.9.The problem areas and the problem influencers

5.4.8. Predict the problem test areas: Dimension 4.

See fig.5.4.8 Problem test areas prediction

1. Software functional & robustness
2. Product performance
3. Compatibility & Interoperability
4. Default and state of art

We have to put special attention in the testplans covering these areas because they are the major influencers on the low scores for the problem blocks.

CONCLUSION → for the whole product creation process with subcontracting.
The areas to improve or to plan an extra effort before starting the project are:

1. The optical interface	**4%**	**1. Software functional & robustness**
2. The SW interface	**20%**	**2. Product performance**
3. AF VCM	**6%**	**3. Compatibility & Interoperability**
4. Interaction SW Companion	**7%**	
5. Installation	**4%**	
6. All Software coverage	**6%**	

Workshop 5.4.8:   Analyze the Problem test area prediction for actions

5.4.9. Predict the problem influencers.

See fig.5.4.9. The problem areas and the problem influencers

We create relations between all blocks. With the relation "1" we earn the value of that block, by putting a "2" we earn "the block value/2" and by putting a 3 we earn "the block value/3". It is possible that "none problem areas" will show up as major "problem area influencers". The "Features-Suite" is not a problem area but it has an influence on many problem areas and as such it has to be considered as an important component.

Conclusion for the problem influencers:

The problem areas:

1. The UVC-OEM webcam companion: the Software and its drivers
2. The integration between software and video card // the Auto Focus algorithm.
3. The basic support of the global PC market with its standard drivers.

The problem influencers:

1. The specialized driver has the biggest impact on many problem areas.

2. The support of this specialized driver in the software and videocard.
3. The Auto Focus algorithm
4. The support of the global PC market with its diversity in drivers.

Workshop 5.4.9:   Analyze the Problem area influencers.

This conclusion
is
the result of
months of work and analysis
of
several teams.
Company managers
should use
this precious data
to
improve.

5.4.9. Profit // loss estimation:

Conventional Product
Creation Process

4D-FMEA product
Creation Process

Higher assured Quality
Assured TTM
Lower cost

No Architectural Detailed Predesign - DFMEA	Milestone Quality (%)	With Architectural Detailed Pre-design - DFMEA 4D-DFMEA	Milestone Quality (%)	Profit & Loss	Profit & Loss justification
x	0%	4D-DFMEA & FPY prediction	95%	-25K$	work + travels 4 weeks
Planning & Programming	50%	Planning & Programming	95%	-2.5K$	work 1day/week
Feasibility	50%	Feasibility	90%	-2.5K$	work 1day/week
Assignment	50%	Assignment	90%	-2.5K$	work 1day/week
Creation-FMEA	50%	Creation	65%	-7.5K$	First test plan execution
+30days	60%	+30days	75%	-2.5K$	work
+30days	70%	Realization	95%	-10K$	Release test plan execution
Realization	80%	MP	Product Q 97%	+20K$	<FCR
MP	Product Q 80%			+200K$	Sales 4 Weeks 5K 10$

Fig.5.4.9_1 Profit - loss calculation based on total product creation guidance and testing.

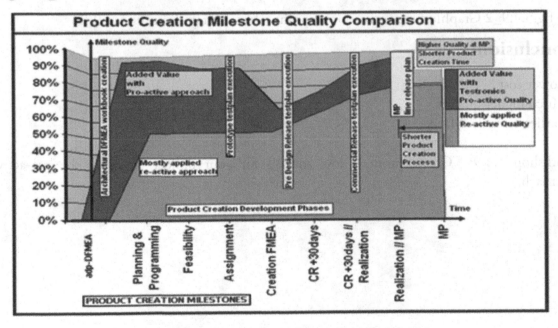

Fig.5.4.9_2 Graphical representation of profit – loss diagram from fig.5.4.9_1

Finally we tried to estimate the profit // loss by being pro-active, by implementing the architectural detailed pre-design DFMEA.

See fig.5.4.9_1 Profit - loss calculation based on total product creation guidance and testing.

The milestones of the conventional product creation process are put on the left side. You can see the milestone quality in red.

The milestones of the pro-active product creation process are put on the right side. You can see the milestone quality in green.
For the traditional product creation process many milestones releases are based on unknown values, like "how did you judge the planning? What was it based on? Did you just take the same time as always?" Because these items are unknown the quality of the milestones can't be good. Therefore we judge 50% because from experience we know that it was even less.

For the pro-active manner we put effort to find evidence for all inputs before starting the process. It means that we have all input. We even use statistics to predict problem areas. At these milestones we know all risk and conditions. Thus the quality of the milestone is very high.

At creation we will see on both sides a lower quality. This is because we do our first tests and have to start debugging. For the pro-active approach we predicted the problem areas and set up already solution scenarios. Therefore we will solve much better, much faster and we will solve all issues in a shorter timeframe.

In the figure 5.4.9_2 you see a graphical representation of the above datasheet.

See fig.5.4.9_2 Graphical representation of profit – loss diagram from fig.5.4.9_1

Conclusion:

1.Lower cost
2.Much higher quality
3.Earlier Time To Market

Workshop 5.4.9: Create profit // loss analysis for your project by using the pro-active approach.

5.5:Product failure over lifetime, from zero hour till "warranty over".

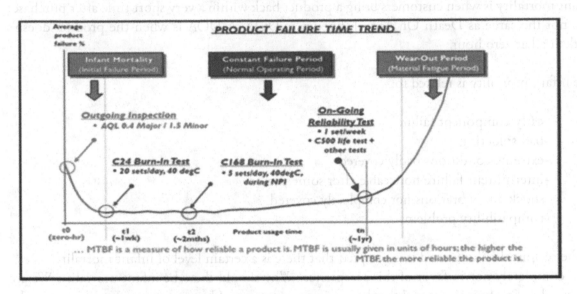

Fig.5.5.1 Product Failure trend is a bathtub shape

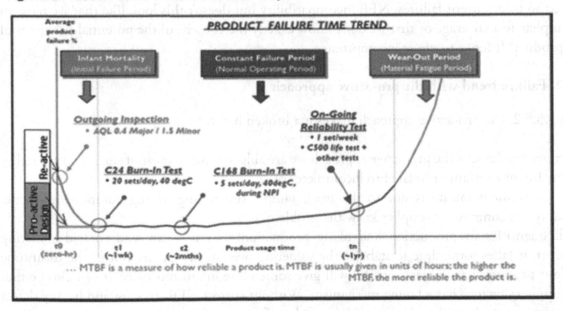

Fig.5.5.2 The pro-active approach makes it a broken bathtub.

5.5.1. Bathtub curve with known processes

See fig.5.5.1 Product Failure trend is a bathtub shape

With the nowadays processes, mostly all re-active, we measure the result of creations.
The infant mortality is also called "child diseases". Most of the diseases a human being gets are in the childhood because we don't have our immune system adapted.
Infant mortality is when customers bring a product back within a very short time after purchase. It is not the same as Death On Arrival, known as DOA. A DOA is when the product doesn't work at all at zero hour.

The infant mortality is related to:

> -early component failure
> -bad soldering
> -extreme conditions badly covered
> -intermittent failure noticeable after some period
> -shock and vibrations not completely covered
> -compatibility problems

In the technical failure trend it is accepted that there is a certain level of infant mortality.
I am completely not in favor of this acceptance. Why should there be infant mortality. We are talking about technical material that has no immune system. I know that many problems can be rerated to intermittent failures, NFF, incompatibility but doesn't this look like that we have an incomplete test coverage or that we don't know exactly the content of the potential problems of our product. It looks to me as incompetence.

5.5.2. Failure trend with the pro-active approach

See fig.5.5.2 The pro-active approach makes it a broken bathtub.

When we implement the pro-active approach we are able to remove most of the infant mortality effects. Infant mortality is related to incomplete work.
Early component failure is due to the small sample size testing during component release. Probably the component supplier knew the problem.
Small quantities in production and many production start ups can lead to bad soldering problem. It takes some time to stabilize the soldering oven in the factory but in the meantime dubious products left the factory. This will give some more infant mortality. It can also be that the supplier's factory has a failing mechanism. Without correct NFF tracking and hourly defect measurement tracking the product quality will be affected.
The extreme conditions badly covered is also a lack of professionalism. When we build in the margins and test requirement we have to be complete and know the end user conditions otherwise we are just looking for trouble. From experience I know that this area is a highly overlooked area and mostly only 90% covered.

The intermittent failure problem is also an incompetence problem. During previous projects no learnings were built up or during previous project no interest was shown in the voice of the customer so that the strange "No Defect Found" or "No Fault Found" issues were not analyzed or they were not worked on. By focusing during the product creation process and during the testplan creation on the possibility of intermittent failure learnings could be build and believe me, testplans can cover intermittent failures. Intermittent failures will introduce No Failure Found problems. These NFFs are a real pain for most companies. I will come back on NFF in the next chapter. The factory can help a lot to assure that the product doesn't contain a real product NFF.

The complete previous chapters all give very good insights on how to approach the product creation process to avoid premature born failures.

Do not

kill

a fly

with

an elephant.

5.6: Do not kill a fly with an elephant

Use appropriate tools and equipment to solve a problem.

I have to introduce this paragraph because it is so bad and so real. It is the real world. Top management decides to correct a parameter. They trust their competences, their black belt, their managers and their employees but they don't challenge the results. What they don't take into consideration is that these managers were responsible for the creation of the problem.

The expression "Don't kill a fly with an elephant" lets us some space for personal interpretations. But what it really means is that we get annoyed by a fly and we use an unrealistic huge over-dimensioned solution to kill it. Costs are irrelevant because the decision comes from very high and thus costs don't matter. We catch an elephant, lift it and drop it on the fly. We use an over dimensioned solution and kill a complete fauna.

But the fly is death too.

The following story is about the quality manager who in contradiction to hierarchical advise, although required by the company project books but neglected by the whole business unit, used his knowledge as six sigma black belt to set up critical parameters for the products, defect control in the factory and reporting after production. In the product creation process he encountered major issues that were waived and passed the commercial release milestone. To improve he worked with helpdesk and returned products to find the major pain points from the customers. However with all his efforts he couldn't improve the product quality as the design wasn't stable enough and too many issues were encountered at design release. Note that he was seen as a "Quality island" because he tried to do his job as real quality manager. Out of this knowledge he developed much later a methodology to assure a reliable and high quality product creation process with a prediction of the problem areas before starting the product creation process.

As you understood already the company had a really lousy customer mindset. The only value was TTM no matter what the quality was.

At a certain moment a KPI, CPP, was introduced to measure how happy the customer was with the purchased product. CPP was deployed to the whole organization. The result was very bad. Master black belt projects were set up, all data was measured. These efforts cost tons of dollars. But it came from the highest management and cost was no point. CPP had to improve no matter what price.

Note also that these results were already known by the above quality manager. He tried creating a real quality system but as "Quality Island" this was impossible. He tried to work pro-active and sense before the problem happens. The system that was setup now was completely re-active and as such new products would always have the first products with lower quality.

Keep in mind during the whole explanation that I am trying to promote pro-activity and use the optimized way of measuring versus re-activity.

5.6.1. The quality manager

Note that when I entered this company I was hired as a quality manager. As such I had to assure that the product creation process was followed, that products were compliant to the project book and Customer Requirement specification. What do you do when you enter a company? You start learning from what the other quality managers did and in the meantime you get involved in some teams. So I was assigned to the webcams and the memory stick group. The webcam group was a very important group in that business unit because it was a high margin product and the sales was quite good. However the return rate was not so good, around 1.5%.

I was interested in what the customers told and as such I read all the calls from the customers to the helpdesk agent. I analyzed the data and tried to get a top 3 pareto. Note that if a quality manager did his work as he should he would have investigated the helpdesk calls and he would have the test results of the returned products. I was the only one doing so. ?????

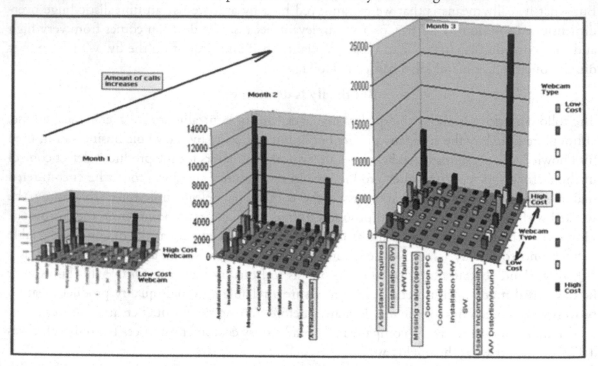

Fig.5.6.1_1 One sheet overview: Three months helpdesk calls

Conclusion:

1.Increasing numbers of calls: 100% increase monthly
2.Major issues:
    1.Highest cost type product is highest problem child
    2.Assistance required, Installation, A/V incompatible, Specifications wrong

I knew that a lot of problems were caused by the Vista OS implementation that year and for the high end product we had many difficulties to get the drivers correct under vista because the drivers for the high end webcams were proprietary drivers, developed by a subcontractor that was not interested in that business anymore.

The products for the roadmap 2008 were already on the market and my frustration was high that I had to commercially release all the products with very severe problems, like that computers had to be too powerful to handle the power needed by the drivers to get high quality video streams with just normal resolutions. The promoted high resolutions would kill the computer performance. Luckily I was able to limit the problem with a driver change.

I monitored intensively the helpdesk calls and I analyzed the written complaints from the customer when they returned a product, all this effort to get an overview of the major problem areas. I didn't need specific problems, I wanted generic areas to work on, like audio, video, drivers, installation. Critics will say that the outcome is not based on correct and detailed numbers. That's right but the outcome confirms the experience I had during design release and now I could back it up with some logical thinking.

This method uses data from 2 different sources, helpdesk and returns, and merges the data in an inventive way to come to the 2 major problem areas, video, drivers and driver installation.

The below figure shows a summary of reported issues to the helpdesk.

Call	Type3	Type5	Type7	Type1	Type2
audio n	88	190	222	470	0
audio q	0	0	0	470	0
audio reglage	0	0	0	157	0
cpu load	0	63	0	0	0
driver	572	507	1553	1097	702
fps 2pics	0	190	222	470	1053
hang	0	317	0	784	702
info	0	63	666	0	1053
install	88	127	444	470	1404
linux	0	63	222	0	351
mac	0	63	0	0	1053
msn	0	190	666	313	0
property n	0	63	666	0	0
skype n	0	127	0	157	0
snapshot n	0	63	222	0	0
video n	44	0	666	784	0
video q	0	507	222	0	1053
video pink	0	127	888	0	0
change bg	0	0	222	0	0
xp sp1	0	63	222	470	351
wifi	0	0	222	0	0
cdrom	44	0	0	470	0
crash	0	0	0	313	351

Calls	Type3	Type5	Type7	Type1	Type2
Audio	88	190	222	1097	0
Driver	704	1078	2663	3134	3159
video	44	888	2441	1254	2106
info	0	571	1997	940	2808
Total	836	2726	7324	6425	8073
	3,29%	10,74%	28,85%	25,31%	31,80%

Call	Type3	Type5	Type7	Type1	Type2
Audio	0%	1%	1%	4%	0%
Driver	3%	4%	10%	12%	12%
video	0%	3%	10%	5%	8%
info	0%	2%	8%	4%	11%

Fig.5.6.1_2 The helpdesk calls summarized into 4 categories, Audio, drivers, video, info.

Webcam	5fps	bild kein	bild qualitat	empty blankpp	erkannt nicht	installt nicht	keine function	mic kein	mic qualitat	sturtzt ab
Type1	0	0	157	470	470	157	0	157	0	157
Type2	0	702	0	0	0	0	0	351	0	0
Type3	0	221	133	44	0	0	0	44	44	44
Type4	0	381	0	0	0	0	381	0	0	763
Type5	0	63	254	127	63	63	0	63	0	0
Type6	193	385	385	193	578	385	0	193	0	0
Type7	0	444	888	444	222	0	0	0	0	222
	193	2197	1816	1278	1333	605	381	808	44	1186
5fps	bild kein	bild qualitat	emptyblankpp	erkannt nicht	installt nicht	keine function	mic kein	mic qualitat	sturtzt ab
1,96%	22,33%	18,45%	12,98%	13,55%	6,15%	3,88%	8,21%	0,45%	12,05%

QRL	Type3	Type5	Type7	Type1	Type2
Audio	88	256	0	157	351
Driver	1188	1089	444	784	0
video	735	1280	1332	157	702
empty	44	319	444	470	0
	2056	2944	2219	1567	1053
	21%	30%	23%	16%	11%

QRL	Type3	Type5	Type7	Type1	Type2
Audio	1%	3%	0%	2%	4%
Driver	12%	11%	5%	8%	0%
video	7%	13%	14%	2%	7%
empty	0%	3%	5%	5%	0%
NFF 75%					
	21%	30%	23%	16%	11%

Fig.5.6.1_3 The QRL analysis

See fig.5.6.1_2 The helpdesk calls summarized into 4 categories, Audio, drivers, video, info.

To get the above data I had to read all the calls and put them into specific categories and then connect them to one of the 4 major problem areas.
Ex. Audio n (audio no), audio q (audio quality) ⊠ Audio category.
I also got returns, the first 100 I always investigated them myself. This process was called the Quick Return Loop (QRL). It meant that it was the first possible moment to get physical returns and test the failure or get a No Failure Found (NFF). Note that the written customer failure was for me a real failure. That I couldn't reproduce resulted in a NFF % but for me it meant that the customer wasn't happy and wrote this message down. For the QRL the below figure gives the numbers.

See fig.5.6.1_3 The QRL analysis

During the investigation I noticed around 80% of all the returned products was NFF. A logical merging would be to take 20% for the QRL because these should be real defects and 80% for NFF and relate them to the calls and calculate the average to find the importance of the category. Note that this number doesn't have to be precise, it has to give an indication were we have to work on.

QRL	Type3	Type5	Type7	Type2	Type1	
Audio	88	256	0	157	351	
Driver	1188	1089	444	784	0	
video	735	1280	1332	157	702	
empty	44	319	444	470	0	
	2056	2944	2219	1567	1053	9840
	21%	30%	23%	16%	11%	

QRL	Type3	Type5	Type7	Type2	Type1
Audio	0,009	0,026	0,000	0,016	0,036
Driver	0,121	0,111	0,045	0,080	0,000
video	0,075	0,130	0,135	0,016	0,071
empty	0,004	0,032	0,045	0,048	0,000

QRL	Type3	Type5	Type7	Type2	Type1	20%
Audio	0,9%	2,6%	0,0%	1,6%	3,6%	
Driver	12,1%	11,1%	4,5%	8,0%	0,0%	
video	7,5%	13,0%	13,5%	1,6%	7,1%	
empty	0,4%	3,2%	4,5%	4,8%	0,0%	

QRL *0,2+ Call*0,8	Type3	Type5	Type7	Type2	Type1
Audio	0%	1%	1%	4%	1%
Driver	5%	6%	9%	11%	10%
video	2%	5%	10%	4%	8%
empty	0%	2%	7%	4%	9%

Call	Type3	Type5	Type7	Type2	Type1	80%
Audio	0,3%	0,7%	0,9%	4,3%	0,0%	
Driver	2,8%	4,2%	10,5%	12,3%	12,4%	
video	0,2%	3,5%	9,6%	4,9%	8,3%	
info	0,0%	2,2%	7,9%	3,7%	11,1%	

| | | | | | | | 7% | 15% | 28% | 23% | 28% |

Fig.5.6.1_4 Merging Calls(fig.5.6.1_2) and QRL(fig.5.6.1_3) into 1 number.

Fig.5.6.1_5 CODN → Cost Of Doing Nothing

See fig.5.6.1_4 Merging Calls(fig.5.6.1_2) and QRL(fig.5.6.1_3) into 1 number.

Conclusion: the major areas to work on were drivers and video and with drivers we could include that the installation of the drivers was also a problem area.

Most issues that I read in the helpdesk calls I knew them from design release and I mostly was overruled by higher management to approve the design release anyway.

I knew now the weight of the webcam types and weight for the audio, driver, video, others category, and could map the issues, that I had seen during design release and during the product test execution and that were probably not gone but still intermittently present, to the FCR and related costs.
If we suppose that if no issues would be present that we would have a minimal amount of returns and a minimal amount of calls then we can try to find the major issues to work on from these issues that we know from design release and release testing. See the fig.5.6.1_4. You also in fig.5.6.1_5 a line "Other FCR bias OK FCR". This is the FCR that you can't improve. It is a bias due to unpredictable behavior of people.

Some of you will put questions about the merging of the QRL and calls numbers. It is of the utmost importance that you choose to work on problems. The criteria that you built in are valid, other criteria would be valid too if they are based on logical thinking and common sense.

See fig.5.6.1_5 CODN → Cost Of Doing Nothing

Category Return Weight:	**from the previous investigation**
1K products	
Issues Weight:	**Assignment with 1, 3, 5 to weight for customer importance**
Type3 FCR=0.65	**Type5 FCR=1.1**
Type3 Cost = 7$	**Type5 Cost = 9$**
Margin 2800$	**= 1000x40%x7$**
CODN loss	**= 1000x0.65%x(7$+10$)x(7-2.1)/7= 77.6$**
Normal FCR loss	**= 1000x0.65%x(7$+10$)x(2.1)/7= 35.1$**

With these formulas you will find your way in building the calculation sheet.

We see that type7, type2 and type1 would be the products to focus on and the orange items on the left of the sheet have the highest impact in solution ratio, respective 27, 23, 28 and in total covers 78% of all the return costs.

By using the above methodology we know the areas to focus on and we know the products to focus on. We are not so much interested in the very specific problems because they will be tackled project per project but you have to find the attention areas. We know the specific areas and the focus point, see the left orange high weight problems. Where must I put the most resources or where do we have to put special effort? This is pro-activity. We find problem areas that are always valid. We can use these findings in all products because they are common sense.

The whole idea is that we have to find the issues that have the highest impact on the overall result.

Note however that this company was stuck so much in their way of working that they were not able to change. How much I wanted to focus on certain items it was impossible to change because this change would have an impact on the whole projects and even makes certain project impossible. Creating webcams without proprietary drivers or without proprietary IP was not thinkable. Discussion closed. This IP was brand image. Even if it screwed up the whole product it had to be present. This discussion about these proprietary drivers took already more than 2 years and we were still standing at the some point.

5.6.2. The official Customer Product Promoter Score

The highest of the highest management got the best idea ever. They introduced a customer satisfaction key performance indicator called Customer Product Promotion (CPP). In this company the customer was lost out of sight for several years already. Now the customer was called to ask about his product experience somewhat a couple of weeks after the purchase. This is a fair score. I knew that that this would mean that the company had to make major changes to comply with the customer wishes. I knew that the score would be dramatically bad and it would mean also that the company would have to reorganize the whole operation. And really the result was very bad.

The company put a major effort in place with several master black belt projects. You know that when it comes from very high costs don't matter.

First actions were to measure the helpdesk calls and to call the customers to gather their ideas. Note however that the company had a helpdesk in place to help the customer, they had a Quick Return Loop in place to collect the first returns but nobody used it or interpreted the data because of unprofessionalism.

Anyway the highest management decided to introduce a huge CPP scoring inquiry. A department was created to handle the cooperation between all departments. Black belt projects were assigned in several business units. This cost tons (100K$) of dollars but what has to be done has to be done. Money doesn't matter. A huge investment was done to measure the remarks of the customer and to analyze the data.

Workshops were started to get the project team ideas. I was member of one of these teams and I told them the experience that I had already about the issues from our products. Note that the questionnaire that was asked at the customers was after the products were bought. This is per definition re-active. Note also that the products had a life cycle of 6 months meaning that solving issues within the measured product was quite impossible because it was impossible to solve them before the product was phased out.

Nobody even put the link between development and the score.

They were not able to create a good product from the first time right. If they would create always new products they would always have bad products. Measuring the comments from the customer is ok but it can only be a process to get better ideas or really out of the logical thinking extraordinary problems. It can't be that the company has to get the normal thinking about problems from these analyses.

The purpose was very good and it was a real blessing for the company that finally the message from the end user was listened to. But the cost was so immense to create special software to analyze the calls, to interpret the data and to deploy the actions. The set up cost millions and actually the data was known if the company used all the available data as helpdesk and QRL.

The following general business unit CPP score was sent out. See the reporting that was send to the teams, a summary of all products. We see an overall general very negative score except for some countries but these positive scores were more related to low samples size.

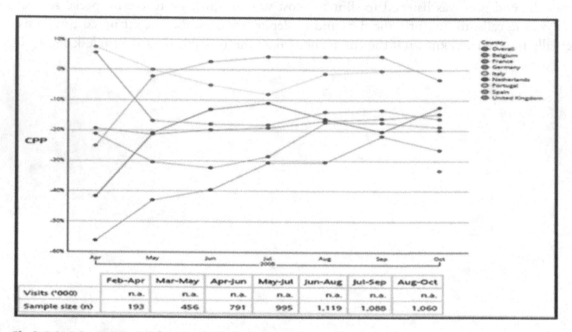

Fig.5.6.2_1 Customer Product Promoter score relation to the countries

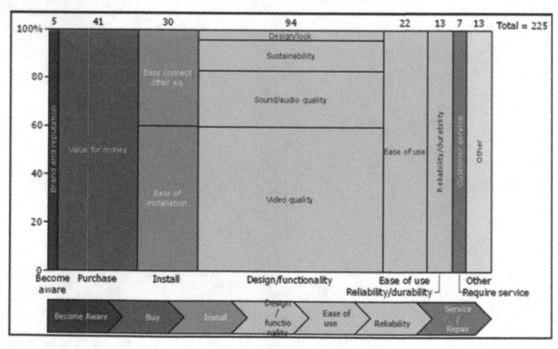

Fig.5.6.2_2 Promoters for positive customer experience

See fig.5.6.2_1 Customer Product Promoter score relation to the countries

We see very scattered scores in the early measurement because of the small sample size. It means that the countries have different product judgment behavior. We must not forget that the sample size can have an influence too as a small sample size doesn't have a statistical importance. The next months there is a merging of the scores to the reality, an overall negative score averaging to -20%. The customers would not promote the products they bought to their friends.

For the company this score was a reflection of the brand index and a negative brand index was a very bad message.

A deeper product questionnaire resulted in an overview as shown below. These generic product main promoters were sent. There was also a monthly email that there was a responsibility to take action on this data.

See fig.5.6.2_2 Promoters for positive customer experience

For this product the main promoters are:

value for money
ease connect
quality
reliability
and service.

From what I knew these were most items that the company was trying to save money on and for my products I improved a lot on reliability by introducing the required performance parameters as critical parameters and MTBF(Mean Time Between Failure). Note that nobody used these parameters anymore although required by the project book, the bible for every project. The monthly datapack that I received from the suppliers was useless as it contained only management data and no customer important data. This was really very bad. I set up a datapack containing the critical parameters and weekly reporting.

It was also like an impossible assignment to make these items happen because I needed the suppliers to do something that was taken for granted for years and even more this "none quality" was taken for granted as "quality". These items were known but due to the product creation setup it was impossible to make it happen.
From the CPP score it was clear that we had to improve video quality, audio quality, ease of use and ease of installation.

The results of these inquiries were quite correct but to come to this conclusion I wouldn't consume so much money. This is pure common sense. This is what everybody wants. If you need to be engineer or master black belt to come to this conclusion and if a company spends tons (100K) of Euros to get this data, my trousers fall on my shoes. This is close to fiction.

Any analysis of the helpdesk calls would give the same result, more detailed but cost zero, the only thing needed was a management concerned about the customer's want and don't wants.

The result of all this data is sent out monthly. It is generalized and actually it doesn't mean anything anymore. What does this data represent in reality?

It represents a number about what customers think about products they bought.

This data comes after the customer bought the product. Does this seem to be re-active? I asked the question during the workshop but it seems that pro-activity doesn't exist in the dictionary. The consumer care and quality management said that solutions would go in the next product. This was the idea it had to go, project team waiting till the products were ready for design release, release testing and finding out that we had crucial failures that we had to accept to comply with Time To Market. This is so re-active.

I told you already that the time to sense a problem would take 5 months. What is the use of this 5 months delayed sensing system if your company is historical inventive and creative? Every new product would have a first of product with bad quality? You have many new products and all these products will have their infant failures. It means that you will just accept the failures in your products?

You will learn in paragraph 5.4 that pro-activity in design can achieve a first time right product.

Workshop 5.6. Summarize the performance KPIs and the costs to gather the data
What do you measure and what is the relation to the customer?

Really I am lost.

Quality is

listening to the customer

and

putting everything in place

to make him happy with our product.

Why do we need problems to act?

Chapter 6
Pro-activity in Mass Production

Why would we introduce pro-activity in mass production? Because we want to know the problems as soon as possible. You know I have been quality manager. I have seen so many times that a quality system is set up with a reaction speed of 5 months. What does this mean? It means that the time between production and the capability of sensing the problem reported by the end user or our company takes 5 months. Let me tell you.

THIS IS NOT A QUALITY SYSTEM.

THIS SYSTEM SUCKS.

However ignorance and stupidity is from this world.

I have known quality managers making it to directors of quality by creating this type of quality. It is hard to believe but it seems that this is the real world.
I have worked in a company that put in place a sensing system without critical parameters, without defect control and reporting monthly for the FPY and solely based on batch inspection. Surprised that epidemic failures were common? I was not. I was surprised that there were not more problems. This quality system sucks. This type of quality system can be set up in an OEM or ODM product creation cooperation if no real customer concerned people are involved.

With pro-activity we can be on a better road to quality.
1. Better and robust development
2. Robust transfer process with factory
3. Use trial and pilot run as DOEs and analyze in depth.

The factory is a huge information lab.

6.1: Product transfer process

The OEM customer creates the CRS, the customer requirement specification, that contains all the requirements the products has to comply to, meaning product technical requirements, software requirements, installation requirements, default requirements, quality requirements, test requirements and all requirements the product has to comply to. This CRS is transferred to the potential supplier and he makes a quote to cover the whole creation and realization process. After getting awarded with the project the supplier will start the product creation process and a little bit desynchronized the supplier will start the product transfer process when he starts creating the prototypes. The factory will be asked to create the prototypes according to a PCBA layout and premature BOM list. For the following processes there is an interaction between factory and development and this process is called the product transfer process.

The product transfer process is the process between the R&D department and the factory to organize the takeover of the product manufacturing. Between these two units, independent units, a process is created to get things passed from one to the other. The development department created a product from which they think it can be mass produced, and they release it to the factory. Not in one step. There are several moments in the transfer process, consisting of a prototype creation and testing, a design release testing and releasing and trial run testing.

The number of defects is a real score parameter of the capabilities of the development department. What we do is just trying to solve the problems. This is not correct. The numbers of issues we encounter are a parameter of the development department quality.
The development department constructs the product according to the specifications created by the architects, product managers and quality managers. It is mandatory that this technology and development department does a real thorough job. They have to create a detailed customer requirement specification. The best would be to use the methodology from chapter 5 to know the problem areas in advance.

The prototype stage:

The prototypes will be tested following the functional test requirements from the supplier. Debugging is done and improvements are made. Certain test plans have to be passed to come to the next milestone and start the production of the first trial run with a higher quantity.
By creating some more prototypes reliability testing could be done with a probability of reliability. Just know that 95% of all issues are so easy to find that you will see most issues here already. If more difficult to trace reliability issues are potential it is in general better to set up specific actions to work on these issues. It is important to detect these as early as possible and it is even more important to know the risks even before the problems happen.

The first trial run:

Now reliability and robustness has to be tested and probably the above mentioned reliability problems will be sensed if the test plan is complete. The critical parameters are tested and the product variance quantified. The whole set of ORT tests is done. When all these tests pass, we can pass the milestone. This is all very standard testing to come to the next factory run, the pilot run.

The pilot run:

The pilot run is the design release for the product. In this stage the packaging has to be complete and the product must be complete, installation CD ready. The full product can be tested against the full product specification.

How can we implement pro-activity in this process? First of all the entire product creation process follows a standard process. But there are ways to do things better. Companies have to adapt to new conditions. This part is going wrong in many companies. Companies have to adapt to changes.

How to adapt to a request from the customer for highest quality? The customer is only loyal to high quality brands. So we have to create these products.

How many times did I encounter that a crucial parameter was not met or was deviating too much from the competition. When the normal benchmarking values are not met the product should be stopped or the component supplier should be contacted to improve.

Selling a product with a quality below the average customer expectation
is killing yourself on long term.

We have to get the products ready BEFORE mass production. Mass production can't be used to tune the product. The product must be ready for mass production. The pilot run should cover the defects at ramp up.

Workshop 6.1: Create the product transfer process with SIPOC
The process to transfer the product to the factory
(prototypes, trial run, pilot run, acceptance)

6.2:Defect control in Mass Production

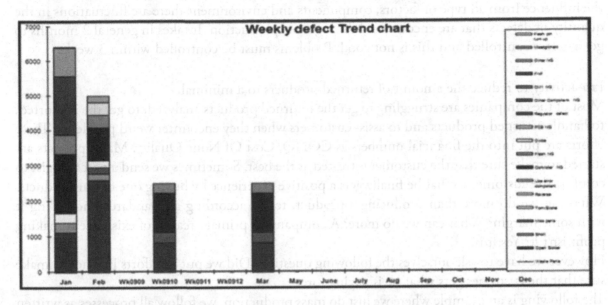

Fig.6.2_1 The defect control sheet

Fig.6.2_2 Raw defect data from the final assembly line

See fig.6.2_1 The defect control sheet

The goal of production is to copy a product compliant to customer's expectations. Because of the influence from all type of factors, components and environment there are fluctuations in the quantity of defects that are encountered during mass production. It takes in general 2 months to get a defect controlled and this is not good. Problems must be controlled within 3 weeks.

Pro-activity to reduce the amount of returned products to a minimal.
Most of the companies are struggling to get the returned products analyzed, to get the FA sorted, to handle returned products and to assist customers when they encounter weird problems. These efforts are put into the financial numbers as CONQ, Cost Of None Quality. Major projects are started to make sure that the customer is treated at the best. Sometimes we send a psychologist to comfort the customer so that he finally gets a positive experience by buying one of our products. What can we do more than producing a product, test it according all standards and selling it with some margin? What can we do more? A company its primary reason of existence is making profit isn't it? Yes it is.
However we have to ask ourselves the following question. Did we put all efforts in place to make sure that the customer gets what he is paying for?
The following is an example where we just do mass production, we follow all processes as written in the books and we badly fail to assure the product quality because we didn't put the customer priorities in place.

The factory mass production defect control

A huge source of product quality information

Merge end-user eyes into the defect reporting priority

The subcontracted product manufacturing to OEM – ODM suppliers has certain risks. These factories are paid to output as much as possible products according to the Quality plan and the Product description. Between the development department and factory there was a product transfer process and after that process the job is completely in the hands of the factory. The factory is reluctant for asking help from the development department. The experience that I have is that factory will avoid by all means asking help from the development department. They will use screening to screen out the defective products. They try to find a system to sense the defective products. But mostly the screening process is not a 100% reliable system and the result is that defective products leave the factory.
It is the task from a company to produce products for the end user, not for themselves. The problems that are seen in the outside world can for 80% be seen in the factory. Would it be a good idea to use the factory as a huge defect reproducing center? A genius setup of certain test criteria will make end user problems visible in factory. We need the WILL to serve our customer. Technically it is possible but do we want to think about the methodology to do it?

Ex.I always base my reasons to do something on examples.

I was marketing support engineer for major OEM customer at a famous brand of CD and DVD burners.

A customer called us with the complaint that during his acceptance testing he saw 10% failure rate during writes on a certain DVD+R media. We had already released this drive to another very important customer and as such we had mass production for this drive. It should be impossible that a 10% failure rate at customer side is not detected at factory. Why do we test at factory if we don't sense this type of failure? We have to test the critical parameters related to the end user. If a 10% failure on a certain media is not a critical parameter to test, then I don't know what we have to test.

To elaborate a little bit about the final assembly, this is final line where a product gets assembled and tested to see if all functionality is working. A CD – DVD burner has to be tested on a lot of media, CDR, CDRW(Lo, Hi, US), DVD-R, DVD+R, DVD-RW, DVD+RW, DVD-DL and within all these there are 3 or 4 write methods, called write strategies. It means that final assembly has to test most of these media. We had that customer complaint about the write failure. The typical error code reporting from factory is written in the next sheet. As you can see it is very difficult to read.

In factory it is a standard way of working that the line engineers have a weekly meeting to discuss about the top5 failures to start improvements.

See fFig.6.2_2 Raw defect data from the final assembly line

The above sheet shows all the detailed failures in low level failure code format. Note that you only see 3 station defects CDRW ultra, DVD-RW and DVD-R. However there are 12 different media and thus also 12 different stations all testing the functionality for every specific media. The defect codes are reported but if development has to read these low level failure data reports they will never be interested in this data because they need 1 day to get anything suitable out of this data.

We set up a more detailed analysis in factory. The first thing to analyze is the defect types and error codes that are seen in the factory with regards to hourly defects and classified in a more user code message type reporting. The fig.6.2_3 was a first approach.

Disc type			
CD-R	8	213	96,24%
CD-RW	27	205	86,83%
DVD-RW	1	178	99,44%
DVD-R	10	177	94,35%
DVD+R	3	167	98,20%
DVD+R D	11	164	93,29%
Flashin	0	153	100,00%
		Total	62,96%

	Error	Defected by hourly							Daily Error code pareto										
		8-9	9-10	10-11	11-13	13-14	14-15	15-16	Set write speed fail (3)	Wrong Disc type (7)	Read error (17)	Read error LBA (19)	Compare fail (20)	Jitter OS (21)	Reserve track fail (25)	Close session fail (26)	Disc Full (29)	Drive Time Out (32)	write rate below spec (34)
LPC	2322							1											
	2324	1			1		1												
	0000		1	1			1												
	2325			4			1												
6813	032	1	1		1													4	
	029	2															2		
	007				3					5									
DVD-RW	007			3						3									
	026			1		1										2			
	025			1											1				
	017				1						1								
CD-R	025		1												1				
	003			1					1										
	026				2		2								4				
	032				1													1	
CDR	007	1							1										
	017			4	0	1	1				15								
	026			1											1				
	025			2	1	3	3	1							10				
DVD-RW	017				1						1								
DVD-R	026		1	2		1									5				
	020			1								1							
	017				1						2				1				
	021				1														
	019				1									1					
DVD+R	017					1					2								
	026					1									1				
DVD+R DL	017		1					1			2								
	026			1	1	2		2							6				
	024		1			1	1												3
	Total	4	7	15	31	11	12	6	1	9	23	1	1	1	12	19	2	5	3

Fig.6.2_3 Hourly defects summary with additional summing information

Fig.6.2_4 End user interpreted and corrected failure report

108

See fig.6.2_3 Hourly defects summary with additional summing information

We interpret the failures from a different perspective. First of all we use normal descriptive sentences to write the type of failure and we add all related failure codes together and use the total column to decide about the highest priority.
For this sheet it seems that the read error has the highest priority and it looks as we did the correct job. However we didn't.

We didn't take into the account the probability of occurrence. Let me tell you a simple example.
When a disc is mounted in the drive the motor has always a probability to fail. We have 12 stations and at each station we run the disc. At the end of the line there were 12 probabilities that the motor fails. When we want to see a write error on a specific disc, for example DVD-R than we have only 1 potential failure. From end user perspective he will complain if he sees 1 failure for writing the whole disc. In factory we write 500MB on a complete disc of 4700MB. The probability from end user perspective is 10 times smaller and this also per station. Most factories write only 100MB which is a perfect method because the factory is not a reliability center, except in ORT. The final assembly in factory is a line that copies products and checks all functional and specification requirements.
The failures have to be prioritized for the end user importance if we want to be a company that is interested in the customer. The write error for a specific media will never be worked on because its probability of occurrence is much too low compared with other failures.

See a more end user optimized representation of the factory failures in fig.6.2_4.
We introduced all customer related values like media importance, market size, media price and occurrence probability.

See fig.6.2_4 End user interpreted and corrected failure report

The important lines are "Type Priority after correction" and "Media Priority". From this sheet the developer can very fast conclude that he should look into Write failures and this for DVD-RW, DVD-R and CDRW-US. Next in failures priority are the functional failures and this for the CDROM media, not the DVDROM media. Do you see how much information is derived from this summary? Probably the designer will find out that the media have related write strategies (timing and laser power) and for the Rom media he will also know why there is a difference between DVDROM and CDROM. For him this sheet is very useful and he gets this data every day in this type of representation. Is this an added value for a company? Yes it is and it is very much. Companies should use this data very preciously. It is for free and very useful. We only need a computer.

Conclusion:

We have to create a useful information channel between factory and development to be pro-active in mass production. We measure the end user important failures and get them in the TOP5. The normal factory processes will take care of the rest during their normal process meetings.

Workshop 6.2: Create a defect control sheet or analyze a defect control sheet
    Analyze the defects and their relations with technical properties
    Create the customer important error messages
    Create the connections between the factory failures and customer failures
    Analyze the occurrence probability at customer and factory
    Create the final reporting sheet for development

6.3:Critical parameters

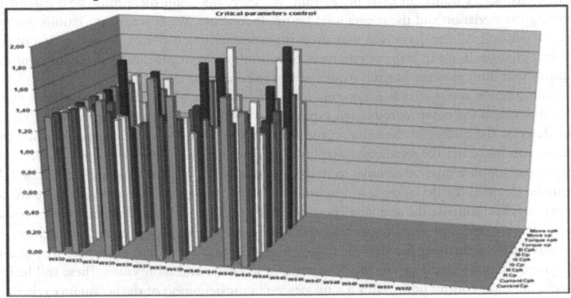

Fig.6.3 Critical parameter distribution chart in Cpk and Cp

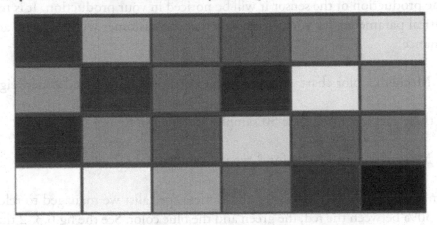

Fig.6.3_1 Macbeth Color chart with an average equal Red, Green and Blue weight.

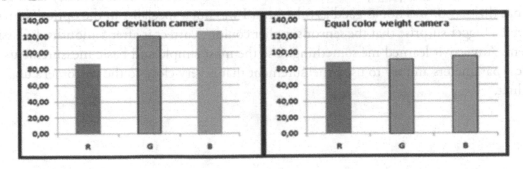

Fig.6.3_2 Measured color fidelity for 2 type of cameras.

See fig.6.3 Critical parameter distribution chart in Cpk and Cp

The compliance to the customer wants are mandatory requirements. We have to check compliance to the "customer's wants" in mass production. For every deviation there must be a parameter showing the deviation and the trigger level for product failure. What are the deviations that we have to control? All the deviations that affect the customer in a way that his perception of the product changes. I am in favor of two types of control, the controls that stop the line and the controls that don't stop the line but require action to correct the deviation.

A small example about critical parameters for a product. One of the critical parameters of an optical device is the color reproduction expressed in RGB values. There are many other values that describe the quality of the color reproduction but for a customer it is important that the picture you make has the same color weights as the original. As customer analysis of the color fidelity the supplier used 50 samples to make a picture of the Macbeth color checker chart. If we make sure that the deviation is within limits for these 50 samples we know that there are no deviations that influence the customer's perception of the camera quality. The next trick is to find out what is the deviation that the customer will notice and where will we put our limit. Therefore we introduce a color filter between the camera and chart and try to find out where we put our acceptance limit by changing the filter's strength to less color and more color. These will be the upper and lower control limits. This is a major step in the definition of the acceptance criteria. Putting a Cpk deviation control on these colors will keep the deviation within limits. If something happens at the production of the sensor it will be noticed in your production. It is mandatory to get all the critical parameters for your product so that the customer will not sense any deviation of its performance.

See fig.6.3_1 Macbeth Color chart with an average equal Red, Green and Blue weight.

The result of these pictures from these 50 samples is shown in fig.6.3_2

See fig.6.3_2 Measured color fidelity for 2 type of cameras.

However even with an architect and color and camera specialist we managed to release cameras that deviated 30% between the red, the green and the blue color. See the fig.6.3_2 the left side. It was by using this customer perception measurement that we found the deviation. My feeling was always that this sensor was not as good as others but I was not a specialist in the optical domain. It was even a bigger surprise that the sensor supplier couldn't correct it after 5 months discussion with them. Anyway it learned me something. Use the most simple and basic measurements for the critical parameters and try to use a measurement that is very close to the customer's usage of the product.

Workshop 6.3: What are your product's critical parameters? Makes sure it is complete. Make sure you measure the parameters that affect the customer's perception.

Factory process controls

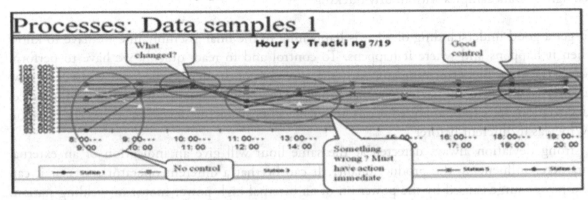

Fig.6.4 Data samples with hourly tracking

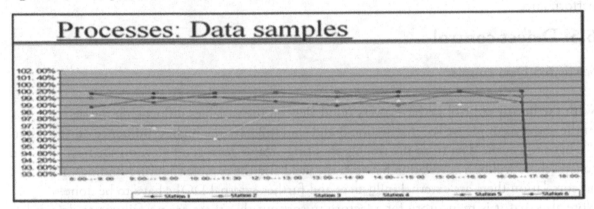

Fig.6.5 Data samples

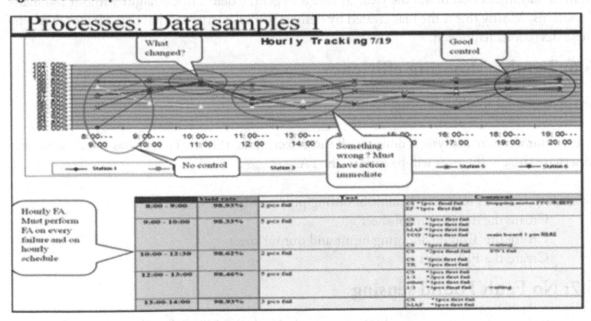

Fig.6.6 Total trouble shooting control

113

6.4: First Pass Yield

See fig.6.4 Data samples with hourly tracking

To get a good understanding of what is happening in the final assembly line we have to know when it happens and where it happens. To control and to react quickly we have to perform hourly tracking. By performing an hourly measurement of the defects the factory can get a view on internal and external parameters that influence the product quality or the line performance. All deviations have to be analyzed. Constant yield values are ok but deviation creates the question "Why was it different before?"

A strong deviation always detected at the same hour will give an indication of an external parameter influencing the product quality. It can be that a different operator starts or it can be a power surge caused by the power on of an external high power motor or welding station. Without hourly control the factory runs a little bit blind and they will never be able to get a best performance.

6.5: Defect control

See fig.6.5 Data samples

With the hourly control we have a very good view on the variables influencing the defects.
In the above example we see that station 3 and station2 have the highest influence on the result. We have to get the data from the trouble shooting area to get a good view on the next steps. In general there will be a correlation between the yield and the trouble shooting findings. If there is no correlation there are external influences and further external DOEs have to be done.
Note also that it is mandatory for a factory to sort out that there are no failures or defects left from development that biases the yield. A rule is to get the defect line straight without too many deviations. Mostly this is the bias created by design. With deviation we have to ask ourselves why these deviations happened.

6.6: Trouble shooting efficiency

See fig.6.6 Total trouble shooting control

Every hour a failure analysis is done on all the defects. Like this the factory has to run smoothly otherwise these influences will be seen in the FA.

Workshop 6.4-6: Pro-active Trouble Shooting process with SIPOC
    Create the defect control sheet
    Create the Trouble Shooting input and output sheet
    Create the FPY sheet

6.7: No Fault Found sensing

How many companies are struggling to get their No Failure Found under control? From my experience I know that this is one of the biggest money eaters from customer care. Black belt projects are started and the whole world is moved for finally "no result".

What is NFF (No Fault Found)?

The customer buys a product and for some reason brings the product back. Our repair center tests and they can't find any problem. This problem has some relation to pro-activity. Because we didn't use all the data from our processes we can't even tell if it is really product related or the rest of the world related. It would help if we knew that the product doesn't have NFF defects incorporated, isn't it?

I have seen major testplans set up to reproduce a NFF returned product. By definition it will be a huge work because there is no failure or it is a low occurring intermittent event.

No Fault found is one of the highest failures that are seen in returned products. A company can fail the expectations of the customer just by products that fail from time to time, occasionally, intermittently. Imagine that you buy an external harddrive. You plug the drive to your computer for making a backup. This is something that most people do at night. The backup is made and the computer will go into hibernation. In the morning you wake up your computer and it doesn't wake up. This is a common problem for USB devices, that they block the bus so that the computer hangs. The host gives a wakeup signal but the device doesn't acknowledge. The computer continues to wait and this hangs a system. The only way to get out of this condition is to restart the system after a power off. If you know that 75% of all the returned products are No Fault Found, we can't reproduce the problem. Now some people start to discuss if we are talking about no defect found, no fault found, can't duplicate or defect not reproducible. It is the last meaning that we are talking about. We get a product returned from a customer or from the final assembly line and we can't reproduce the problem that the customer or line employee encountered. It doesn't mean that the product has a real problem. The problem can be related to the other connected products or the equipment problem in the factory.

How can you get a standardized procedure to tackle no fault found? The problem with no fault found is that you will not see the problem otherwise it would not be named no fault found.

How to tackle this problem in a structured way? We don't know if the product has a "build in" not reproducible intermittent problem. Where can we sense a "build in" none reproducible problem? In the factory thousands of products are produced every day. This is the place to sense the problem. Even if the NFF probability is only 0.5% you will see 15 events on 3000 units (mostly 1 shift). But the factory has to be set up to sense the NFF. This means that the final assembly line has to be a first pass yield line. Every single none defect has to be put in the repair bin to be repaired by the trouble shooter. The trouble shooter has a major task to sense the NFF. When a trouble shooter gets a problem reported and he can't reproduce the problem the problem has to be located in the equipment or in the product. He can eventually try to reproduce the problem at the failing station. It is major to register the problem relations to equipment. Equipment can never be the problem. They must be repaired. If a certain station is the reason for defects that station has to be repaired. After repair of the equipment it must be the product that has a "build in" no fault found. When storing and analyzing these no fault founds we can get a better view on when the problem occurs. We can even create a special firmware to store the last internal technical hardware events just before the problem occurred. The developer can derive useful information from these last 2 error codes. If we connect even some low hardware level information to it he will be very happy and will find your NFF fast. Or by analyzing the pare to of highest occurrences we can find the best way for a solution. The test in factory on high volume is the best validation of the solution and believe me it is the only one that works.

When using this system we are sure that the hardware doesn't have a "build in" no failure found failure. It means also that if we get returned product with no failure found we know that it has to be related to the external factors, systems and interfering material.

NFF re-active versus pro-active

1.Investigate the RMA returned drives: Re-active mode

- We need the failing symptom
- Screening is very time consuming and very costly.
- Solutions take time to implement with potential production loss.
- Very slow solution process: boat transport 4 weeks, sales channel 4 weeks, return product 4 weeks, FA ready 4 weeks, failure solution 4 weeks = total 5 months defects
- Very bad customer performance.
- Solution is months too late.

2.Use the factory as information lab: Pro-active mode

Factory failures: capture rate is not 100% → 10% failures go to customers
- Factory(SMT/FA) component failures
    - > 1000DPPM must get 8D(1 WK).
- Factory(SMT/FA) solder failures
    - > 1000DPPM must get 8D(1 WK).
- Supplier to component suppliers' communication must be much more knowledgeable.
- Because of the early sensing we have some time to solve the problem.
- We can avoid the screening

Analysis from the factory

Failure symptom resulting in NFF.

1.3.1.A or B or C related.
1.3.2.System: Product access and recognition related.
1.3.3.Power consumption and heat related.
1.3.4.Mechanical related.

NFF symptom	Pareto %	Category	
A can't be read	12	Support related	Support Read
No read	7		Read
No mount	17		Mounting
Can not access	30	System	System/power on
Not recognized	6		System/power on
Can't eject/insert	25	Power / Heat	Component
Noise	3	Mechanical	ME

Fig.6.7_1 Factory measurements related to NFF defects

Product instability investigation

Factory Functional Station False Rejects

Nearly all failures are False Reject.
-this FR is an accepted disease.
-FR is a failure and must be analyzed.

Time	Yield Rate	Input Qty	Production line DATA	Rejected (Total)	Error type /NGQty	Remark	026	Fail Rate (026)	017	Fail Rate (017)
8:00 - 9:00	100,00%	81		0	0		0		0	
9:00 - 10:00	97,69%	130		3	026 1pcs 017 1pcs 019 1pcs		1		1	
10:00 - 11:00	98,46%	129		2	026 2pcs		2		0	
11:00 - 13:00	96,24%	147		7	026 4pcs 017 3pcs		4		3	
13:00 - 14:00	99,16%	118		1	026 1pcs		1		0	
14:00 - 15:00	94,74%	114		6	026 3pcs 017 3pcs		3		3	
15:00 - 16:00	95,33%	107		5	026 3pcs 017 2pcs		3		2	
16:00 - 17:00	96,55%	139		6	026 3pcs		3		3	

Fig.6.7_2 False rejects an accepted disease

Product instability investigation
Factory 5 times 100 drives.

Unknown Products	True Failures	High Freq failur	ailures after 5 passe	First test Rejects	2nd test rejects	3rd test reject	4th test rejects	5th test rejects
100%	4%	11%	40%	19%	16%	16%	18%	17%

Fig.6.7_3 100 drives 5 times retry all stations from final assembly

117

See fig.6.7_1 Factory measurements related to NFF defects

As we are trying to measure NFF in the factory we have to be aware that factory can encounter 2 types of NFF.

1. The equipment failure

2. The product failure

To get the NFF sorted out in factory and to make it measurable the factory has to set up a system to sense the equipment failures. If equipment fails it can't be counted as a NFF. This equipment failure must have a trigger value to get the maintenance crew repair the equipment.

One mandatory requirement has to be put in place in the final assembly line. The line operator can't retest the failure in the line. Every failure must have as consequence that the failed product goes into the repair bin. This repair bin goes to the trouble shooting area and these engineers are the only qualified engineers to judge if the failure is a real failure. They will retry and mark the product with a sticker if the failure isn't repeatable. They will also store the failure in a database. If the number of none repeatable failures is too high on a station they will request the maintenance crew to repair the station.

They will also keep a database of the type of none repeatable failures. With the system of saving the last 2 failures in an eeprom we can easily find the internal hardware problem.

Other experiments, General information and Must knows to understand that NFF:

See fig.6.7_2 False rejects an accepted disease

From the above shift defect report we see that we have 1 to 7 defects in an hour but they are all not reproducible. This calls for further action with a real DOE with a GR&R.

Drive instability test according to GR&R process

We try to reproduce the NFF with 100 products and we do a 5 times repeat test. Note that we test all 12 stations for every single test and note all failures. Therefore it is possible that the failure rate is much higher.

DOE: Factory 100 drives rerun 5 times trial.

See fig.6.7_3 100 drives 5 times retry all stations from final assembly

In the above sheet we can see several things:
1. The real production failures
2. The high frequent failures are 11%
3. The number of failures after all 5 tests on all 12 stations = 40%
4. The number of 1th test, 2nd test till 5th test rejects
5. The number of NFFs = 20%

These results are very bad because there are too many failures and the NFF rate is outrageous. The whole message is only that NFF can be investigated if we will do our best and if we use the data from the factory because here we will find a huge amount of failure codes here.

Summary: Six instability data reports including 1 DOE with 100 drives.

Product instability investigation.

NFF Summary	Total NFF	Power On/Off or No Post	Mounting	Writing/ Reading	ME mov/Led	Comments
1.RMA Aging // reliability only	99	10	14	73	2	1.Servo problem during mounting(power on and illegal medium) and writing 2.Reading fails immediately. 3.Writing fails mostly all at the first cycle.
2.USA Warehouse error codes	23	3	4	16		1.Support1 error code 1 2.Support2 error code 2 3.Support3 error code 3 4.No support -> a mounting problem. 5.Communication failure
3.B2B customer 10 products analysis at RMA	10	1		8	1	1.One product has a marginal behavior for low power or specific power slew rise time on 5Volt. 2.Two supports solid read failures 3.The other products have all a possibility to fail on support2. The failing media must be retested pass on a golden sample product othe
4.Twelve products analysis at RMA	12		7	5		1.The customer reported failures are visible in the eeprom but we were not able to reproduce the problem. 2."No post returned" looks to me as a power on problem. Can be a real power reset problem but most probably it is a product/system combination problem.
5.Analysis local	21	7	4	7	3	1.product not seen in setup 2.Support N issues.
6.Control Run 100 products repeatability NFF	100	NA	NA	NA	NA	1. 20% of all products will give a NFF failure. 2. The testing methodology reduces even the occurance rate, meaning that in reality the failure rate is bigger.

Fig.6.7_4 Summary of all instability investigations

Action to reduce NFF.

Support related (cannot read Support)

CSD: Customer Support Division (RMA) will pass support knowledge of eeprom to R&D to focus on the returned media problems.

MY0P92257016956S01KZ	Support Type :	Support 1	Support Type :	No Support
Customer Complaint: CAN'T READ	Media Format :	Data	Media Format :	Unknow
	Failure Item :	N/A (0xFE)	Failure Item :	N/A (0xFE)
	Failure Reason :	N/A (0x00)	Failure Reason :	N/A (0x21)
	Failure Speed :	No Information	Failure Speed :	No Information
	Failure Temperature :	0x31	Failure Temperature :	0x24
	Failure Address :	0x000000	Failure Address :	0x000000
	Information : 97:26:05 76:80:80		ATIP/ADIP Information : 0:0:0:0:0:0	
	Last test 2 Commands :	0x28 0x25	Last test 2 Commands :	0x25 0x46
	Sense Code :	02 04 00	Sense Code :	02 24 00

Fig.6.7_5 Detailed failure codes in eeprom

NFF failures at Testhouse

Drives	Failure	No data	No disc	DVD+R9	DVD+R	DVD-R	DVD+RW	DVD-RW	RW-Lo	CCR	Cdrom	RW-US	RW-U S4
1.1					MCC003 7:49								
1.2									973422/744380 1 792710/744180 3				
1.3					MCC003 7:49	MCC03R020 1 037302					040903 2		
1.4					MCC003 7:49								
1.5		023A00			MCC003 7:49	MCC03R020 2 037302							
1.6					MCC003 7:49	MCC03R020 3 037302							
1.7				MHM001 2 037201	MCC003 7:49					972060/79597 4			
1.8					MCC003 7:49		OPTODISCOV60 030C00						
1.9							RICOHJPNW01 1 031100						
1.10	No Post		023A00 3 040900 1										
1.11			023A00						972060/755060 1 020400				
1.12			037S02 2	MHM001 1 037302									
1.13												793425/744380 2 040903 052100	
Total	**23**		4	2	7	3	1	1	2	1	1	1	

Fig.6.7_6 Failure reporting for a number of drives

See fig.6.7_4 Summary of all instability investigations

We need the will to set up the product for DFM (Design For Manufacturing for repair) so that we can analyze also low frequent failing errors. How can we still learn something after a none reproducible failure happened? We created a firmware that stores the latest 2 error codes in an eeprom. We store also related deep hardware codes to know in which state the microcontroller or the hardware was at that moment.
See in fig.6.7_5 a representation of the failure codes.

Actions to reduce NFF:

1.Support related symptoms.

2.System: product access and recognition symptoms.

3.Power consumption and heat related.

4.Mechanical related.

5.Specific artificial NFF reducing actions

1.Support related (cannot read support)

See fig.6.7_5 Detailed failure codes in eeprom

With these failures codes the developer can find out what the controller was doing at the failing moment. By analyzing enough failures he will see correlation failure reasons and he can find potential solutions. He can know if the media is the cause and adjust the write strategy.

Also the media can be the cause of the problem, which deviates in its support to a normal write strategy.

See fig.6.7_6 Failure reporting for a number of drives

With the above table we can see which media have an influence because the media can be a major influencer for NFF.

These data are found from returned drives but this method is re-active and as such not an example to follow but only to use if we failed on all stages. When we talk about returned drives we have drives that a customer returned and was not happy about his purchase. This return of products has to be limited to its minimum. We have to use the opportunity to use the factory data to create a stable product.

2.System related

-product access and recognition symptoms. The system problem related should not have a high impact as the drive was created for OEM major computer manufacturers and we have a detailed list of the computers. The drives could be intensively tested on the target systems.
This type was the lowest priority and as the risk of return for this type of problem was quite small.

3.Power Consumption and heat related:

-Components at the edge of their specification: -design review with thermal and power profile request.
-Burning or failing components.

4.Mechanical related problems:

-Noise problems // LED // Tray related problems.
> 1.Action: Monitor very closely the component defects in factory.
>> Second source components introduced with same specification but failing more easily after all.
>> Ex. Voltage regulator change from 2 to 1 chip.
> 2.Action: Monitor Computer Manufacturer Aging station Failure to verify the 50C influence.
>> Monitor and use the information. (see previous aging station sheet).
> 3. Factory SMT must reduce its failure output to Final Assembly to ZERO.
>> SMT: 100% capture rate "Bad contacts and bad soldering".
>> Ex. Screening for a tray problem: bad soldering on resistor array.
>> The "Final Assembly functional stations -component and -SMT related failures" must be reduced to zero.
> 4.All Component failures must be handled with 8D.

5.Specific artificial NFF reducing actions.

This has nothing to do with pro-activity. This is just trying to reduce the negative effect of failures at the customer's premises.
-put supplier's engineer to screen the failing drives in the customer factory to avoid the discussion about the NFF between customer and supplier. This method is used when a subcontracting manufacturer assembles for example the complete computer and sub assemblies show issues at their assembly process. To avoid discussion an acceptance test is created and after this test the product is accepted. This is more an agreement to continue rather than a solution and according to me it has not much to do with quality.

NFF: Preferred way NFF

The preferred way is that we use the failed drives in factory and analyze the failure codes. This is how it finally was set up. In the factory we have to avoid the False Reject.

In many factories and even companies "False Reject" is a "Final Assembly" accepted disease.

We set up a system to avoid false rejects and retest in line in the factory so that we could analyze the failure codes. Because of the huge amount of data we could very fast find correlations and solve the real NFF problem.
We realized now that the factory was NFF free, except for some equipment problem once in a while. We also found that quite some media were creating NFF . So we adapted the write strategy and timing so that this type of media coverage was much better. We could now finally say that the returned drives with NFF were real NFF drives. The drive didn't suffer hardware wise from NFF but it were real outside world NFF issues. The amount of NFF was already decreased to 10% which is an acceptable NFF level as customers can also return products without any reason.

Overall conclusion: Pro-activity by using the factory as information source.

1.70% of all our NFF and Fires can be avoided by intelligent use of the factory.

2.NFF is a real failure. (drives should be returned to the supplier).

3.Use factory as data collection – information source.

In general NFF is a management problem, not providing enough knowledge and experience to find correct solutions. This is the reason why many NFF projects fail.

Workshop 6.7. Create NFF analysis for your products
        Procure the trouble shooting flow from factory
        Draw the equipment failure process
        Draw the product NFF process
        Factory, Development, Customer care

Chapter 7
Pro-activity in Product Portfolio Creation.

It is not my goal to write too much about the product portfolio creation but it is an essential part of a healthy company to have a pro-active approach in their portfolio creation. You saw in paragraph 5.1_1 that there is a process " Ideation Strategy Option Creation" that should output tasks for idea creation. To be healthy a company needs 30% revenue from high margin products, innovative products.

There are companies solely busy with benchmarking with the competition, with speed racing, with low profit products. It means that this company's result depend on others and on very small market movement. You understand that these companies are very fragile and minor changes in the market will cause major problems, negative cashflow. I knew a memory stick company that was fully dependent of the flash price and the flash suppliers. For every price change they took risks. If the trend of the price is always in the same direction than it is a controllable business but the flash price was completely unreliable. This supplier was our major ODM subcontractor for our memory sticks and they had a very unpleasant habit, always making mistakes and never delivering flawless products. They were not busy neither with pleasing the customer. They were busy with doing business.

This company wanted to think about new products but forgot that innovative products have to differentiate by something, not just higher capacity, not just a new type of LED. It meant also that this company was not able to create 15% of innovative products and they were all years struggling with profit.

7.1: High margin products versus low margin products.

What is low margin and what is high margin?

In production the normal margin is around 3% on asset value. If we count a product turnaround of 4 months they get a 9% margin on year basis. I would call this low margin.

Big companies need in general higher margin to finance their overhead, their fixes assets, their top management and a lot of blunders. The high margin is around 40% on year bases. However this margin leads sometimes to confusion. To calculate the feasibility the product manager calculates the margin on the product price. This is not the correct method. He should take the product turnaround time into account. What is the product turnaround time? It is the time between the payment of your suppliers and the payment of the bill from your customers. The margin will be made on this time. It means also that on year basis the total margin will be much higher. If the product turnaround time would be 4 months the year margin would be 120%. Giving the wrong interpretation to the meaning of margin can kill your product price because you should only count 13% margin in your feasibility study.

Don't tell me that this type of mistake doesn't exist, I have worked in a team where this was the margin rule. It is useless to tell that we always had huge difficulties to meet the market price.

Product turnaround time and effect on margin.

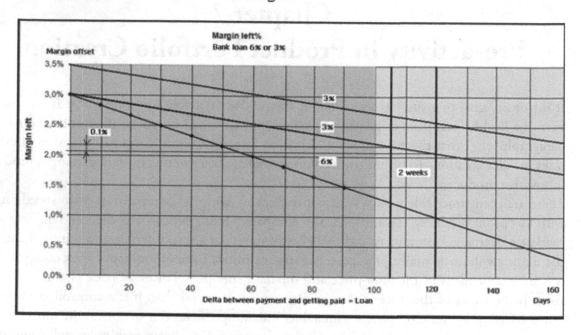

Fig.7.1 Margin versus product turnaround

Although the margin is such an important number some companies neglect it and accept everything from their customers regarding payment terms. Conclude from the above graph that a reduction of 2 weeks in turnaround time gives you 0.1% and if you have a 3 month turnaround time you will even win 0.4% in total.

It is not the goal of this book to give financial advice but I thought that the above would be good to know.

7.2: Lateral leadership

Lateral leadership: a management that promotes creativity, innovation from the employees so that the whole company is involved in new ideas. Imagine the brainpower that a company disposes off. Use it to create new ideas.

How can an individual manager or team, of managers acquire the characteristics and skills of the lateral leader? Some traits of the lateral leader come naturally to certain people, whereas others find them difficult to acquire. However, if you buy into the benefits of a lateral leadership style, it is possible for you to change your actions and behavior to become more of a lateral leader than you are today. The recommended initial action is a two-day off-site workshop – the lateral leadership course.

You have to find out what type of people are working in your company because it is a mandatory requirement that you use all types of thinking to get most ideas, to get a highest quantity of ideas.

7.2.1. Workshop to categorize the employees in "type of brains".

Brain characteristics
1. Self-organize input into patterns
2. Learn from the past
3. Small steps
4. Very good at making connections

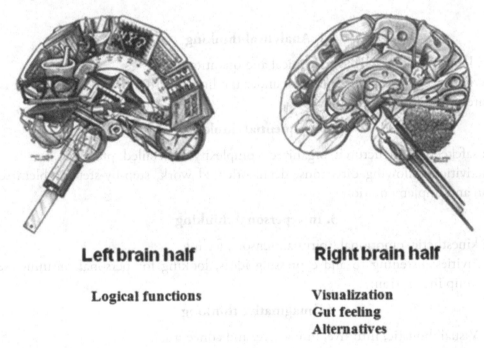

Left brain half

Logical functions

Right brain half

Visualization
Gut feeling
Alternatives

Fig.7.2.1 The "type of brains".

Mostly we rely on the left brain half:
1. Use known patterns for solving problems
2. Requires less energy
⇨ Familiar feeling ⇨ We feel comfortable ⇨ But this is not creative

How to get pulled out of the comfort zone? Lateral Thinking Definition

"A way of thinking which seeks solutions to intractable problems through unorthodox methods or elements that would normally be ignored by logical thinking". Lateral thinking paves a casual way of releasing and directing creative energy, a structured method to create and elaborate ideas and create consciousness how the brain can be used.

Use all type of brains

But there is a second condition to get the workshops most successful, you have to get the 4 different types of people together using the MBTI indicator or the HBDI indicator.

The MBTI indicators will give you:

1.the "rational" (judging) functions like thinking and feeling

2.the "irrational" (perceiving) functions like sensing and intuition.

The HBDI instrument will give you:

1. Analytical thinking

Key words : logical, factual, critical, technical and quantitative.
Preferred activities: collecting data, analysis, understanding how things work, judging ideas based on facts, criteria and logical reasoning.

2. Sequential thinking

Key words: safekeeping, structured, organized, complexity or detailed, planned.
Preferred activities: following directions, detail oriented work, step-by-step problem solving, organization and implementation.

3. Interpersonal thinking

Key words: kinesthetic, emotional, spiritual, sensory, feeling.
Preferred activities: listening to and expressing ideas, looking for personal meaning, sensory input, and group interaction.

4. Imaginative thinking

Key words: Visual, holistic, intuitive, innovative, and conceptual.
Preferred activities: Looking at the big picture, taking initiative, challenging assumptions, visuals, metaphoric thinking, creative problem solving, long term thinking.

If you just try to imagine how you are you will easily be able to put a score for all the 4 types of thinking. Look at the preferred activities to find your score. Make sure you have a homogeneous team with at least all brain types present.

7.2.2. The company's vision and its components

If 'the vision and strategic objectives are not already agreed and have to be hammered out during the course you have to make sure first to have this straightened out.

There are various well established methods and approaches to setting the vision, and indeed it is a topic worthy of a book and a workshop all its own. The facilitator leads a discussion, using creative techniques to define the vision. Our preferred approach is to start by examining four elements that underpin the vision:

-a purpose;
-a mission;
-a culture;
-a set of values;

The purpose is the fundamental reason for the organization's existence. To define this you need to ask at the most basic level. Why do we exist and what is it we do?' The Coca-Cola purpose is expressed in what they call their promise:

> The Coca-Cola Company exists to benefit and refresh everyone who is touched by our business.

The mission is a forward-looking expression of the purpose of the company. It should inspire and challenge without being too prescriptive. So the Coca-Cola mission is:

> When we bring refreshment, value, joy and fun to our stakeholders, then we successfully nurture and protect our brands, particularly Coca-Cola. That is the key to fulfilling our ultimate obligation to provide consistently attractive returns to the owners of our business.

This feeds into their vision, which is:

> To become the world's leading consumer company for automotive products and services.

The mission should be simple, clear and meaningful to staff and customers alike.

The corporate culture expresses the style and manner in which the company operates. So the culture statement generally makes great play of employee empowerment, development, challenge and so on. Adjectives such as innovative, enthusiastic, dynamic, energetic, customer-focused, learning, decentralized and empowered are often used. Unfortunately the cultures that the directors use are often some distance away from the reality in the organization. The culture statement should not be a wish list for the future. You should spend time in a workshop session defining as much about the current corporate structure as you can - strengths and weaknesses. Then look realistically at how the culture could be strengthened and developed. The resulting culture statement should be rooted in reality, but with sensible goals for improvement.

Similarly with the values. If the current values are to cut costs and maximize profitability, it is no good mouthing pious declarations about customer service, employee development and environmental responsibility. The workshop is a good place to take a hard look at the current corporate values and to build on the good ones. The values statement should be a summary of what the organization really stands for, what it believes in and aspires to.

Finally the vision is a short statement that encapsulates the essence of the culture, values, purpose and mission. The four components underpin the vision, but the vision cannot simply amalgamate all the previous statements, or it becomes unwieldy and impractical. Brevity in vision statements is a great benefit

If time allows, the group should examine each of the four components, of the vision, using creative questioning approaches and techniques such as six thinking hats to analyze and evaluate different options. Finally the vision is a short statement itself that should be discussed and agreed. It is also possible to start with the vision and then analyze the four components subsequently. Either way it is important to agree a statement that is succinct and has a strong sense of purpose and direction. It will be the platform for the strategy and change objectives that follow.

The objectives of a strategy workshop should be:

1.Agree the vision, goals and strategic direction (if this has not already been done).
2.Set a common agenda for change and for lateral leadership.
3.Set creativity goals and metrics.
4.Develop skills in questioning, challenging assumptions, creative problem solving, idea generation, idea analysis and evaluation.
5.Agree action plans to implement a lateral leadership program throughout the organization.

Note: the six thinking hats are:
1.the white hat: the information hat. What more information do you need from the idea?
2.the red hat: the emotions hat. What do you feel with the proposed idea?
3.the yellow hat: the optimism hat. Everyone has to say good about the proposal.
4.the black hat: the pessimism hat. Everyone has to find fault with the idea.
5.the green hat.: the growth and possibilities hat. How can the idea be improved?
6.the blue hat: the process hat. Is the process working well?

COMMUNICATIONS PLAN

Communication of the vision and its components, and of the strategic plan, is so important that it is worth spending a section of the workshop on this topic. The challenges to be addressed depend on where the organization is in its development and communication. These are the sorts of topics that should be considered:

-How can we communicate the vision, the mission, culture, values and purpose to the internal audience? Which of these messages should we prioritize?
-How can we communicate the vision to the external audience?
-How can we get the staff to buy into the vision?
-How can we develop departmental and personal objectives in line with our strategic goals?

-How can we communicate the need for creativity, innovation, ideas and entrepreneurial spirit?
-What mechanisms, training or processes can we put in place to generate a flow of ideas from all the staff?

This should lead to a list of actions to improve communication, ranging from the mundane to the truly radical.

OBJECTIVES

One of the sets of objectives that should be defined is the list of innovation objectives. This should support the strategic objectives and define targets such as:

-the number of new products;
-revenue from new products in existing markets;
-revenue from new markets or ventures;
-the number of new strategic partnerships;
-the areas where new processes or procedures will be implemented;
-a target for the number of prototypes entering the selection funnel;
-a target for the number of ideas generated by the staff;
-how long it takes for new products to go from idea approval to launch?

7.2.3. Idea Creation and Innovation.

A company needs differentiating products to keep face to continental or even global changes. Therefore a company needs a process or unit that is specialized in the creation of ideas. I don't say that they have to create the ideas. No. They are responsible to take the initiative. They have to follow the market inputs, the market changes, the demand changes, the economical inflation or deflation situation, the economical prosperity areas and the growing continents. With all these inputs in their ideation process they have to start investigations to find the future markets and the future products. It is also this team that should give guidance in killing product lines or just keeping a small best of a kind product. This knowledge is also called the SWOT and PEST analysis and it is the entire basis for acceptance or rejected the ideas that come from workshops.

If we think now about the idea creation as such we talk about the workshops to generate ideas. In many companies the higher management steers this input to the company. They find new products during a golf session or a pub visit. This is really not the preferred way.

The company has to get a process in place to get the following done:

-lead the group through the various phases;
-encourage their full participation;
-motivate them to believe in their own creative potential;
-challenge them to achieve more than they thought possible;
-teach them about lateral leadership;
-develop their creative and leadership skills.

BRAINSTORMING

The problems to be solved should be general rather than specific to the organization. Typical topics that can be used for the exercises are:

-How can we get everyone to use public transport?

-How could we win all the gold medals at the next Olympic Games?

-How can we get everyone to take more exercise?

-How can we persuade young people not to start smoking?

-How could you double the sales of your local flower shop?

The facilitator encourages all the participants to contribute, and rigorously enforces the rule about no early judgment or criticism of ideas.

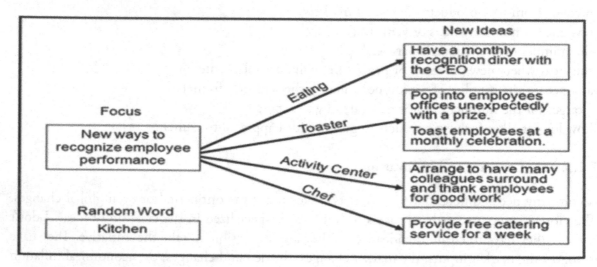

Fig.7.2.3_1 Example of focus question.

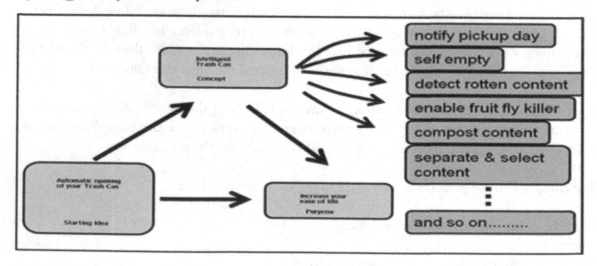

Fig.7.2.3_2 Generating more ideas

CREATIVE EXERCISES

It is a good idea to intersperse creative leadership exercises with specific business-oriented activities. The variety helps stimulate the brain, and the lessons learnt from the exercises can immediately be put to use on the business tasks. The facilitator will choose one or two from the following - the choice depends on timing and suitability for the group.

How to create new ideas?

Random entry	→ use dictionary words
Break the rules	→ reverse the good and the bad idea.
The worst solution	→ what is the worst solution?
Idea cards	→ round table ideas
Found objects	→ random objects

See fig.7.2.3_1 Example of focus question.

The random entry is one of the basic and best idea creation methods. Formulate the main question to find ideas. What is the focus? "New ways to recognize employee performance?"
Find a random word in a dictionary and start working on the relations between the word and the focus idea.

Generation of more ideas:

See fig.7.2.3_2 Generating more ideas

From the "starting idea" we retrieve a "concept" to fulfill the "purpose".
From the concept we generate more ideas. (the green ideas)

The number of ideas is the most important parameter and can be used as a metrics for the idea creation process success.

Different ideas to create ideas: Innovation metrics

7.2.4. Harvesting.

The Harvesting phase includes:
- **New concepts**
- **Interesting concepts**
- **Looking for relationships between ideas**

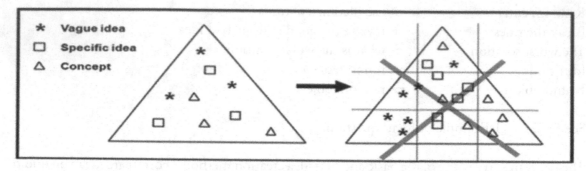

Fig.7.2.4_1 Looking for relationships

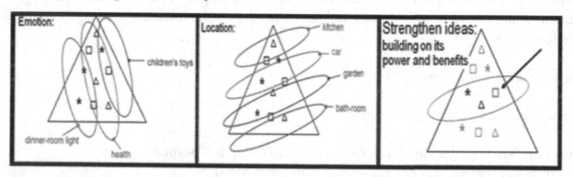

Fig.7.2.4_2 Shaping ideas by adding constraints

	Kitchen	School	Factory	Car	Garden	Bathroom	Livingroom	Children toy	Health	Dining room light	Total
Idea 1	1										1
Idea 2	1				1				1		2
Idea 3	1	1		1	1				1	1	10
Idea 4	1	1	1	1					1		6
Idea 5	1		1	1					1		4
Idea 6	1			1	1				1	1	5
Idea 7				1	1						2
Idea 8				1							1
Idea 9				1							
Idea 10				1						1	3
Idea 57	1								1	1	8
Idea 58	1	1		1					1	1	5
Idea 59			1	1							2
Idea 60				1							1
Total	25	10	3	15					45	7	

Fig.7.2.4_3 Putting weights to categories and ideas.

From this moment we will try to find similarities in the ideas. What is common to the ideas? Can we use higher domains, relations like food, emotion, medical, low cost or any other similarity to gather more ideas together. It is possible to create an excel sheet with all the type of categories that you want to discuss or summarize.

See fig.7.2.4_1 Looking for relationships

See fig.7.2.4_2 Shaping ideas by adding constraints

See fig.7.2.4_3 Putting weights to categories and ideas.

The creation of a value on different categories is useful to have an idea in which direction to think. Also the type of idea that fits most in all categories is good to know. But from this point onwards it depends on the creation team to distil workable ideas. In our example from fig.7.2.4_3 we can think that "Kitchen" , "Health" are the categories to work on. Furthermore the "idea 2" and "idea57" are the ideas to work on because they match in most of the customers categories.

Don't forget that your management has to give approval about the ideas. Do not start such actions from below in the hierarchical chain. Make sure you have your management support before the start. You can have the idea but let the manager think that the idea was his.

Workshop 7.2.1-4:   Ideation and Innovation
                      Type of brains
                      Vision and its components
                      Idea generation, concept, harvesting.

Fig.7.2.5. Management logic

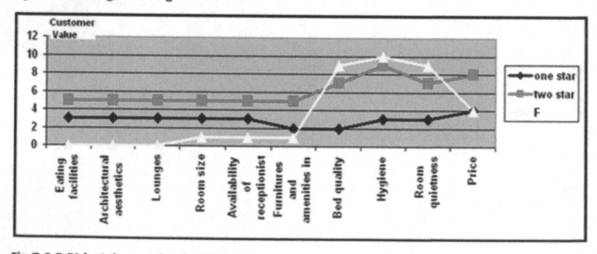

Fig.7.2.5.F1 hotel example of added value.

7.2.5. Value Innovation.

The economical crisis overwhelms all papers and banks are threatened with bankruptcy. Even though there is a rough weather in the financial situations, some companies will come out of the crisis stronger.

Why is this?

Several potential scenarios:

1. The first one is that the company had enough resources before the crisis and didn't get damaged too much.
2. The other scenario is that the company was driven with a value innovating vision.

What do we have to do to comply to the market needs? What can the customer afford? What will be the materials that are needed? Shouldn't we get more combined products or shouldn't we find out how we can offer value to the needs of certain market groups?

Ex: F hotels

Market Intelligence proved that a certain market needed cheap hotel prices with a high level of comfort, bed quality, hygiene, room quietness. The F hotel business was a success by focusing on what the customer needed.

See fig.7.2.5.F1 hotel example of added value.

An economical crisis has always an overreacting effect on symptoms. However the symptoms have to be corrected. In the meantime people tend to spend less money but they will still spend money. If your company creates value that is worthwhile spending money for you are a winner. Apparently you are not with what you are doing now. Don't blame it too much on the circumstances. Act.

Value Innovation can help you passing the difficult times.

Chapter 8
Pro-activity conclusion.

At the end of this book it is good to put some generic ideas that I wanted to promote to you.

1.Pro-activity in Business processes

Measure and quantify the inputs of all business processes

2.Pro-activity in Product Creation

Add knowledge, experience and statistics in a multi dimensional concept FMEA.

3.Pro-activity in Idea and Value creation

Idea creation to differentiate and assure a future

I hope I helped you to take a certain distance from all aligned processes that lead to none conformance to customer's expectations. When the customer's expectations aren't met I hope that you take this book and try to find a way to reshape your ideas and processes to get them aligned with the customer.

Your customer's expectation is your food.

Eat it or you deserve to die.

Make the difference for your customer.

Win.

Chapter 9
Six Sigma and Deming.

A Dr. W. Edwards Deming's principles support the global success of Toyota, Proctor & Gamble, Ritz Carlton, Harley-Davidson, and many other leading organizations. His teachings are essential for the effective application of Six Sigma, Lean Manufacturing, Loyalty/Net Promoter and other quality improvement, customer retention and business growth methods.

Dr. Deming's profound, yet simple, success strategies offer your organization a proven system to achieve lasting growth and success. The principles apply universally to business, healthcare, education—in fact, to any enterprise—and to you personally.

I want to promote Dr. Deming's idea's at the end of this books as it are guidelines that I support through my whole book. This is a proven method how the world can work better. However, for many reasons, many companies aren't able to adopt the ideas.

9.1: Deming's 7 deadly diseases

1.Lack of constancy of purpose in planning product and service that will have a market, keep the company in business, and provide jobs.
2. Emphasis on short term profits: short-term thinking (just the opposite of constancy or purpose to stay in the business), fed by fear of unfriendly takeover and by push from bankers and owners for dividends.
3. Evaluation of performance, merit rating, or annual review.
4. Mobility of management: job hopping.
5. Management by use only of visible figures, with lithe or no consideration of figures that are unknown or unknowable.
6. Excessive medical costs.
7. Excessive cost of liability, swelled by lawyers who work on contingency fees.

9.2: Deming's 17 original points for management

1."Create constancy of purpose toward improvement of product and service, with the aim to become competitive and to stay in business and to provide jobs."

For the company that wants to stay in business, the two general types of problems that exist are the problems of today and the problems or tomorrow. It is easy to become wrapped up in problems of today but the problems of the future demand, first and foremost, constancy of purpose and dedication to keep the company alive. Decisions need to be made to cultivate innovation, fund research and education, and improve the product design and service, remembering that the Customer is the most important part of the production line.

2."Adopt the new philosophy, we are in a new economic age. Western management must awaken to the challenge, must learn their responsibilities, and take on leadership for change."
Government regulations and antitrust activities need to be changed to support the well-being of people. Commonly accepted levels of mistakes and defects can no longer be tolerated. People must receive effective training so that they understand their job and also understand that they should not be afraid to ask for assistance when it is needed. Supervision must be adequate and effective. Management must be rooted in the company and must not job-hop between positions within a company.

3."Cease dependence on inspection to achieve quality. Eliminate the need for inspection on a mass basis by building quality into the product in the first place."

Inspection is too late, ineffective, and costly when the product leaves the door. Quality comes not from inspection when the product leaves the door. Quality comes not from inspection, scrap, downgrading, and rework on the process.

4."End the practice of awarding business on the basis of price tag. Instead, minimize total cost. Move toward a single supplier for any one item, on a long-term relationship of loyalty and trust."

Price and quality go hand in hand. Trying to drive down the price of anything purchased without regard to quality and service can drive good suppliers and good service out of business. Single-source suppliers are desirable for many reasons. For example, a single-source supplier can become innovative and develop an economy in the production process that can only result from a long-term relationship with its purchaser. Lot-to-lot variability within a one-supplier process is often enough to disrupt the purchaser's process. Only additional variation can be expected with two suppliers. To qualify a supplier as a source for parts in a manufacturing process, perhaps it is better first to discard manuals that may have been used as guidelines by unqualified examiners to rate suppliers. Instead suppliers could be asked to present evidence of active involvement of management encouraging the application of many of the S4/IEE concepts discussed in this book. Special note should be given to the methodology used for continual process improvement.

5."Improve constantly and forever the system of production and service, to improve quality and productivity, and thus constantly decrease costs."

There is a need for constant improvement in test methods and a better understanding of how the customer uses and methods and a better understanding of how the customer uses and misuses a product. In the past, American companies have often worried about meeting specifications, while the Japanese have worried out uniformity, i.e., reducing variation about the nominal value, continual process improvement can take many forms. For example, never-ending improvement in the manufacturing process means that work must be done continually with suppliers to improve their processes. It is Important to note that, like depending on inspection, putting out fires is not a process improvement.

6."Institute training on the job."

Management needs training to learn all aspects of the company, from incoming materials to customer needs, including the impact that process variation has on what is done within the company. Management must understand the problems the worker has in performing his or her tasks satisfactorily. A large obstacle exists in training and leadership when there are flexible standards for acceptable work. The standard may often be most dependent on whether a foreperson is having difficulty in meeting a daily production quota. It should be noted that money and time spent will be ineffective unless the inhibitors to good work are removed.

7."Institute leadership, The aim of supervision should be to help people and machines and gadgets to do a better job. Supervision of management is in need of overhaul, as well as supervision of production workers"

Management should lead, not supervise. Leaders must know the work that they supervise. They must be empowered and directed to communicate and to act on conditions that need correction. They must learn to fix the process, not react to every fault as if it were a special-cause problem, which can lead to a higher defect rate.

8."Drive out fear, so that everyone may work effectively for the company."

No one can give his/her best performance unless he/she feels secure. Employees should not be afraid to express their ideas or ask questions. Fear can take many forms resulting in impaired performance and padded figures. Industries should embrace new knowledge because it can yield better job performance. Not be fearful of this knowledge because it could disclose some of our failings.

9."Break down barriers between departments. People in research, design, sales, and production must work as a team to foresee problems of production and in use that may be encountered with the product service."

Teamwork is needed throughout the company. Everyone in design, sales, manufacturing, etc. can be doing superb work, and yet the company can be failing. Why? Functional areas are sub optimizing their own work and not working as a team for the company. Many types of problems can occur when communication is poor. For example, service personnel working with customers know a great deal about their products, but there is often no routine procedure for disseminating this information.

10."Eliminate slogans, exhortations, and targets for the work force asking for zero defects and new levels of productivity. Such exhortations only create adversary relationships, as the bulk of the causes of low quality and low productivity belong to the system and thus lie beyond the power of the work force."

Exhortations, posters, targets, and slogans are directed at the wrong people, causing general frustration and resentment. Posters and charts do not consider the fact that most troubles comes from the basic process. Management needs to learn that its main responsibility should be to improve the process and any special causes for defects found by statistical methods. Goals need to be set by an individual for the individual, but numerical goals set for other people without a roadmap to reach the objective have an opposite effect.

11a."Eliminate work standards (quotas) on the factory floor. Substitute leadership."

Never-ending improvement is incompatible with a quota. Work standards, incentive pay, rates, and piecework are manifestations of management's lack of understanding, which leads to inappropriate supervision. Pride of workmanship needs to be encouraged, while the quota system needs to be eliminated. Whenever work standards are replaced with leadership, quality and productivity increase substantially and people are happier on their jobs.

11b."Eliminate management by objective. Eliminate management by numbers, numerical goals. Substitute leadership."

Goals such as "improve productivity by 4% next year" without a method are a burlesque. The data tracking these targets are often questionable. Moreover, a natural fluctuation in the right direction is often interpreted as success, while small fluctuation in the opposite direction causes a scurry for explanations. If the process is stable, a goal is not necessary because the output level will be what the process produces. A goal beyond the capability performance of the process will not be achieved. A manager must understand the work that is to be done in order to lead and manage the sources for improvement. New managers often short-circuit this process and focus instead on outcome (e.g., getting reports on quality, proportion defective, inventory, sales, and people).

12a."Remove barriers that rob the hourly worker(s) of their right to pride of workmanship. The responsibility of supervisors must be changed from sheer numbers to quality."

In many organizations the hourly worker becomes a commodity. He/she may not even know whether he/she will be working next week. Management can face declining sales and increased costs of almost everything, but it is often helpless in facing the problems of personnel. The establishment of employee involvement and participation plans has been a smokescreen.

Management needs to listen and to correct process problems that are robbing the worker of pride of workmanship.

12b."Remove barriers that rob people in management and in engineering of their right to pride of workmanship. This means, inter alia, abolishment of the annual or merit rating and of managing by objective."

Merit rating rewards people who are doing well in the system; however, it does not reward attempts to improve the system. The performance appraisal erroneously focuses on the end product, not leadership to help people. People who are measured by counting are deprived of pride of workmanship. The indexes for these measurements can be ridiculous. For example, an individual is rated on the number of meetings he or she attends; hence, in negotiating a contract, the worker increases the number of meetings needed to reach a compromise. One can get a good rating for firefighting because the results are visible and quantifiable, while another person only satisfied requirements because he or she did the job right the first time; in other words, mess up your job and correct it later to become a hero. A common fallacy is the supposition that it is possible to rate people by putting them in rank order from last year's performance. There are too many combinations of forces involved: the worker, coworkers, noise, and confusion. Apparent differences in the ranking of personnel will arise almost entirely from these factors in the system. A leader needs to be not a judge but a colleague and counselor who leads and learns with his/her people on a day-to-day basis. In the absence of numerical data, a leader must make subjective judgments when discovering who, if any, of his/her people are outside the system, either on the good or the bad side, or within the system.

13."Institute a vigorous program of education and self-improvement."

An organization needs good people who are improving with education. Management should be encouraging everyone to get additional education and engage in self-improvement.

14."Put everybody in the company to work to accomplish the transformation. The transformation is everybody's job."

Management needs to take action to accomplish the transformation. To do this, first consider that every job and activity is part of a process. A flow diagram breaks a process into stages. Questions then need to be asked about what changes could be made at each stage to improve the effectiveness of other upstream or downstream stages. An organizational structure is needed to guide continual improvement of quality. Statistical process control (SPC) charts are useful to quantify chronic problems and identify sporadic problems. Everyone can be a part of the team effort to improve the input and output of the stages. Everyone on a team has a chance to contribute ideas and plans. A team has an aim and goal toward meeting the needs of the customer.

Vocabulary:

Adp-DFMEA:
Architectural detailed predesign – DFMEA. The pro-active methodology to prevent problems by putting more effort on research before a project start.

AF:
Auto Focus feature from a camera. It means that the camera has a build in mechanism to move the lens to set the focus point optimized.

BG-BU:
Business Group – Business Unit.

Central Limit Theorem:
In probability theory, the central limit theorem (CLT) states conditions under which the sum of a sufficiently large number of independent random variables, each with finite mean and variance, will be approximately normally distributed.

CODN:
Cost Of Doing Nothing.

Cpk:
the process capability index or process capability ratio is a statistical measure of process capability: The ability of a process to produce output within engineering tolerances and specification limits. The concept of process capability only holds meaning for processes that are in a state of statistical control.

$$Cpk = \min \frac{(USL - u)}{3 \times s} , \frac{(u - LSL)}{3 \times s}$$

u = mean
s = sigma
USL = Upper Specification Limit
LSL = Lower Specification Limit

Cpk estimates what the process is capable of producing if the process target is centered between the specification limits. If the process mean is not centered, Cp overestimates process capability. Cpk < 0 if the process mean falls outside of the specification limits. Assumes process output is approximately normally distributed.

CPP:

Customer Product Promoter score. A score retrieved by asking the customer his score about the product he purchased. Would you recommend the product to your friend?

DFSS:

DFSS has the objective of determining the needs of customers and the business, and driving those needs into the product solution so created. DFSS is relevant to the complex system/product synthesis phase, especially in the context of unprecedented system development.

DFMEA:

Design Failure Mode Effect Analysis. What is the effect if you encounter a certain failure? See also FMEA.

DMAIC:

Define the problem, Measure the problem, Analyze the problem, Improve the problem and Control the improvement. DMAIC is the basis for a 6 sigma project.

DR:

Design Release. The milestone to accept the design after release testing. The design release testing consists of functional testing, reliability testing, interoperability testing, compatibility testing and more.

EBIT:

Earnings before interest and tax.

Eeprom:

Electrical erasable programmable read only memory. Eeprom have in general very small capacity and mostly they are serial accessible through I2C bus.

Eight step Discipline:

8D: eight step discipline to solve a problem and make sure it never occurs again.
8D is a formal and meticulous approach to solving complex problems. The 8D process uses a combination of effective techniques and tools to focus a cross functional team through a very

detailed analysis of the problem that has brought them together. When followed diligently, 8D will lead to the discovery of the root causes and possible solutions with consideration of cost, timing, effect on customers, and the impact on the organization.

EMI:
Electromagnetic interference

EPA:
Environmental Protection Agency, Statistical Training Course for Ground-Water Monitoring Data Analysis, EPA/530-R-93-003,

Epidemical FCR:
An epidemical failure is a failure that passed through all the checks and verifications and ended up at the customer side with an effect on the performance, functionality or security so that the products have to be returned, recalled or that an action has to be undertaken to repair or rework the products at the customer's premises.

FA:
Failure analysis. When products are returned for repair a failure analysis report has to be made. What was the real reason for the failure? How can it be avoid? What will be put in place to make the product creation improve? From a FA there can be actions on the company's website to update all the products in the market.

FCR:
Field Call Rate, the product return rate. How many products do the customers bring back?

Firmware:
The software that a microcontroller runs and executes resulting in the execution of his specific task. The firmware from a KBD controller makes sure that the key you pushed is translated to a key command to the computer.

FMEA:
Failure Mode Effect Analysis.
A Failure mode and effects analysis (FMEA) is a procedure for analysis of potential failure modes within a system for the classification by severity or determination of the failure's effect upon the system. It is widely used in the manufacturing industries in various phases of the product life cycle. Failure causes are any errors or defects in process, design, or item especially ones that affect the customer, and can be potential or actual. Effects analysis refers to studying the consequences of those failures.
DFMEA: Design – FMEA
PFMEA: Process – FMEA
CFMEA: Concept – FMEA

FPY:

First Pass Yield. These are the number of products that will pass all production final assembly stations without failure. You have to know that all specifications and customer important specifications have to be tested in final assembly. When a product passes all these tests it complies with the functional requirement.

FA:

Failure Analysis.

FTE:

A Full Time Engineer. An engineer hired full time.

GR&R:

Gage Repeatability& Reproducibility. Is the failure repeatable by any operator? Is the failure reproducible? If I do exactly the same tests will the problem be there too?

HDTV:

High Definition TV

HW:

Hardware, all related to components, printed circuit board.

I2C:

A special protocol for serial devices with low throughput. The protocol has a hardware interface over 2 wires.

KPI:

Key Performance Indicators (KPI) are financial and non-financial measures or metrics used to help an organization define and evaluate how successful it is.

MIP:

Market Introduction Process.

MTBF:

Mean Time Between Failure. Mean time between failures (MTBF) is the arithmetic mean (average) time between failures of a system. The MTBF is typically part of a model that assumes the failed system is immediately repaired (zero elapsed time), as a part of a renewal process.

MTTF:

Measures average time between failures with the modeling assumption that the failed system is not repaired.

NDF:

No Defect Found. During the test no defect is found.

NFF:

No Fault Found. The repair center gets a product and after test the engineer can't find a failure.

RONA:

Return   On   Net   Assets

A measure of financial performance calculated as:

$$= \frac{\text{Net Income}}{\text{Fixed Assets} + \text{Net Working Capital}}$$

The higher the return, the better the profit performance for the company.

NDA:

A non-disclosure agreement (NDA), also known as a confidentiality agreement, confidential disclosure agreement (CDA), proprietary information agreement (PIA), or secrecy agreement, is a legal contract between at least two parties that outlines confidential materials or knowledge the parties wish to share with one another for certain purposes, but wish to restrict access to. It is a contract through which the parties agree not to disclose information covered by the agreement. An NDA creates a confidential relationship between the parties to protect any type of confidential and proprietary information or a trade secret. As such, an NDA protects non-public business information

OEM: An original equipment manufacturer or OEM is typically a company that uses a component made by a second company in its own product, or sells the product of the second company under its own brand.

ODM:

An original design manufacturer (ODM) is a company which designs and manufactures a product which is specified and eventually branded by another firm for sale. Such companies allow the brand firm to produce (either as a supplement or solely) without having to engage in the organization or running of a factory. ODMs have grown in size in recent years and many are now sufficient in size to handle production for multiple clients, often providing a large portion of overall production. A primary attribute of this business model is that the ODM owns and/or designs in-house the products that are branded by the buying firm. This is in contrast to a contract manufacturer (CM).

OEM:

Original equipment manufacturer, the original manufacturer of a component for a product, which may be resold by another company.

OLAP:

An OLAP (Online analytical processing) cube is a data structure that allows fast analysis of data.[1] The arrangement of data into cubes overcomes a limitation of relational databases. Relational databases are not well suited for near instantaneous analysis and display of large amounts of data. Instead, they are better suited for creating records from a series of transactions known as OLTP or On-Line Transaction Processing.

ORT:

On going Reliability Test. The reliability tests like high temperature, temperature change, drop test etc...

PEST analysis:

If appropriate the team can analyze the political, economic, social and technological factors that could influence the business over the next five years. The sorts of questions that are raised are:

-What if there was a revolution in our major overseas market?

-What if there was a change of government and policy in key areas?

-What are the demographic changes affecting our customers?

-What could threaten or change the technology that our products use?

-What if fashions changed radically?

-What if interest rates were to double or triple?

PCP:

Product Creation Process.

Screening:

The process to perform a certain test that should detect the encountered problem so that the factory doesn't release any products that are infected by the problem.

QA:

Quality Assurance.

QFD:

Quality function deployment (QFD) is a "method to transform user demands into design quality, to deploy the functions forming quality, and to deploy methods for achieving the design quality into subsystems and component parts, and ultimately to specific elements of the manufacturing process.", as described by Dr. Yoji Akao, who originally developed QFD in Japan in 1966, when the author combined his work in quality assurance and quality control points with function deployment used in Value Engineering.

QRL:

Quick Return Loop. The process to return as fast as possible the first 100 returned products to have as fast as possible an idea about the problems.

R&D:

Research and Development.

RMA:

A Return Merchandise Authorization or Return Material Authorization (RMA) is a transaction whereby the recipient of a product arranges to return goods to the supplier to have the product repaired or replaced or in order to receive a refund or credit for another product from the same retailer or corporation. In practice, an RMA is only issued after a series of tests.

ROFO:
Rolling Forecast. The prediction of the result, like a weather forecast.

Six Sigma:
Six Sigma is a business management strategy, initially implemented by Motorola that today enjoys widespread application in many sectors of industry. Six Sigma seeks to improve the quality of process outputs by identifying and removing the causes of defects (errors) and variation in manufacturing and business processes. It uses a set of quality management methods, including statistical methods and creates a special infrastructure of people within the organization ("Black Belts" etc.) who are experts in these methods? Each Six Sigma project carried out within an organization follows a defined sequence of steps and has quantified financial targets (cost reduction or profit increase).

SW:
Software. All related to programming of something. This something can be a microcontroller, a mainframe or a simple personal computer.

SWOT analysis:
The team analyses the strengths, weaknesses, opportunities and threats facing the organization. This should be a familiar exercise to most business people. It is a critical review of the company's position in the market. Participants are usually accurate in their assessment of the strengths and weaknesses of the organization, but they do not think widely enough in the opportunities and threats categories.
A threat is anything that can undermine your market position or take customers away. For example, videoconferencing over the internet is a threat to airlines, because businessmen and women might prefer to videoconference with a remote client rather than fly to visit them.
Similarly with opportunities. Who would have thought there was an opportunity for Virgin Group to go into trains or cola drinks? The facilitator should use creative techniques to stretch the imaginations of the participants and encourage them to think outside their normal boundaries.

SIPOC:
Supplier Input Process Output Customer format as process mapping.

Testlab:
Test laboratory

T&D:
Technology and Development

Value Innovation:

Creating value by merging several existing products into 1 product that has a higher customer satisfaction than the separate products. Or creating value by focusing on a specific market segment and providing in that expectation the highest satisfaction.

VCM:

Voice Coil Motor. This is the motor (1 mm) that drives the lens movement for AF.

VOC:

Voice of the customer (VOC) is a term used in business and Information Technology (through ITIL) to describe the process of capturing a customer's requirements.